T0271978

GENETICS, GENOMICS AND BREEDING OF BERRIES

Genetics, Genomics and Breeding of Crop Plants

Series Editor
Chittaranjan Kole
Department of Genetics and Biochemistry
Clemson University
Clemson, SC
USA

Books in this Series:

Published or in Press:
- Jinguo Hu, Gerald Seiler & Chittaranjan Kole: *Sunflower*
- Kristin D. Bilyeu, Milind B. Ratnaparkhe & Chittaranjan Kole: *Soybean*
- Robert Henry & Chittaranjan Kole: *Sugarcane*
- Kevin Folta & Chittaranjan Kole: *Berries*
- Jan Sadowsky & Chittaranjan Kole: *Vegetable Brassicas*
- James M. Bradeen & Chittaranjan Kole: *Potato*
- C.P. Joshi, Stephen DiFazio & Chittaranjan Kole: *Poplar*
- Anne-Françoise Adam-Blondon, José M. Martínez-Zapater & Chittaranjan Kole: *Grapes*
- Christophe Plomion, Jean Bousquet & Chittaranjan Kole: *Conifers*
- Dave Edwards, Jacqueline Batley, Isobel Parkin & Chittaranjan Kole: *Oilseed Brassicas*
- Marcelino Pérez de la Vega, Ana María Torres, José Ignacio Cubero & Chittaranjan Kole: *Cool Season Grain Legumes*

GENETICS, GENOMICS AND BREEDING OF BERRIES

Editors

Kevin M. Folta

Horticultural Sciences Department
Institute for Food and Agriculture Sciences (IFAS)
University of Florida
USA

Chittaranjan Kole

Department of Genetics and Biochemistry
Clemson University
Clemson, SC
USA

CRC Press
Taylor & Francis Group
Boca Raton London New York

CRC Press is an imprint of the
Taylor & Francis Group, an **informa** business
A SCIENCE PUBLISHERS BOOK

First published 2011 by CRC Press

Published 2019 by CRC Press
Taylor & Francis Group
6000 Broken Sound Parkway NW, Suite 300
Boca Raton, FL 33487-2742

ISBN 978-1-57808-707-5 (hbk)

Library of Congress Cataloging-in-Publication Data

Genetics, genomics and breeding of berries / editors: Kevin M. Folta,
Chittaranjan Kole. -- 1st. ed.
        p. cm.
    Includes bibliographical references and index.
    ISBN 978-1-57808-707-5 (hardcover)
1. Berries--Genetics. 2. Berries--Genome mapping. 3.
Berries--Breeding.   I. Folta, Kevin M. II. Kole,
Chittaranjan.
    SB381.G464 2011
    634'.7--dc22

                                                      2010035715

Visit the Taylor & Francis Web site at
http://www.taylorandfrancis.com

and the CRC Press Web site at
http://www.crcpress.com

Preface to the Series

Genetics, genomics and breeding has emerged as three overlapping and complimentary disciplines for comprehensive and fine-scale analysis of plant genomes and their precise and rapid improvement. While genetics and plant breeding have contributed enormously towards several new concepts and strategies for elucidation of plant genes and genomes as well as development of a huge number of crop varieties with desirable traits, genomics has depicted the chemical nature of genes, gene products and genomes and also provided additional resources for crop improvement.

In today's world, teaching, research, funding, regulation and utilization of plant genetics, genomics and breeding essentially require thorough understanding of their components including classical, biochemical, cytological and molecular genetics; and traditional, molecular, transgenic and genomics-assisted breeding. There are several book volumes and reviews available that cover individually or in combination of a few of these components for the major plants or plant groups; and also on the concepts and strategies for these individual components with examples drawn mainly from the major plants. Therefore, we planned to fill an existing gap with individual book volumes dedicated to the leading crop and model plants with comprehensive deliberations on all the classical, advanced and modern concepts of depiction and improvement of genomes. The success stories and limitations in the different plant species, crop or model, must vary; however, we have tried to include a more or less general outline of the contents of the chapters of the volumes to maintain uniformity as far as possible.

Often genetics, genomics and plant breeding and particularly their complimentary and supplementary disciplines are studied and practiced by people who do not have, and reasonably so, the basic understanding of biology of the plants for which they are contributing. A general description of the plants and their botany would surely instill more interest among them on the plant species they are working for and therefore we presented lucid details on the economic and/or academic importance of the plant(s); historical information on geographical origin and distribution; botanical origin and evolution; available germplasms and gene pools, and genetic and cytogenetic stocks as genetic, genomic and breeding resources; and basic information on taxonomy, habit, habitat, morphology, karyotype, ploidy level and genome size, etc.

Classical genetics and traditional breeding have contributed enormously even by employing the phenotype-to-genotype approach. We included detailed descriptions on these classical efforts such as genetic mapping using morphological, cytological and isozyme markers; and achievements of conventional breeding for desirable and against undesirable traits. Employment of the in vitro culture techniques such as micro- and megaspore culture, and somatic mutation and hybridization, has also been enumerated. In addition, an assessment of the achievements and limitations of the basic genetics and conventional breeding efforts has been presented.

It is a hard truth that in many instances we depend too much on a few advanced technologies, we are trained in, for creating and using novel or alien genes but forget the infinite wealth of desirable genes in the indigenous cultivars and wild allied species besides the available germplasms in national and international institutes or centers. Exploring as broad as possible natural genetic diversity not only provides information on availability of target donor genes but also on genetically divergent genotypes, botanical varieties, subspecies, species and even genera to be used as potential parents in crosses to realize optimum genetic polymorphism required for mapping and breeding. Genetic divergence has been evaluated using the available tools at a particular point of time. We included discussions on phenotype-based strategies employing morphological markers, genotype-based strategies employing molecular markers; the statistical procedures utilized; their utilities for evaluation of genetic divergence among genotypes, local landraces, species and genera; and also on the effects of breeding pedigrees and geographical locations on the degree of genetic diversity.

Association mapping using molecular markers is a recent strategy to utilize the natural genetic variability to detect marker-trait association and to validate the genomic locations of genes, particularly those controlling the quantitative traits. Association mapping has been employed effectively in genetic studies in human and other animal models and those have inspired the plant scientists to take advantage of this tool. We included examples of its use and implication in some of the volumes that devote to the plants for which this technique has been successfully employed for assessment of the degree of linkage disequilibrium related to a particular gene or genome, and for germplasm enhancement.

Genetic linkage mapping using molecular markers have been discussed in many books, reviews and book series. However, in this series, genetic mapping has been discussed at length with more elaborations and examples on diverse markers including the anonymous type 2 markers such as RFLPs, RAPDs, AFLPs, etc. and the gene-specific type 1 markers such as EST-SSRs, SNPs, etc.; various mapping populations including F_2, backcross, recombinant inbred, doubled haploid, near-isogenic and pseudotestcross; computer software including MapMaker, JoinMap, etc. used; and different types of genetic maps including preliminary, high-resolution, high-density, saturated, reference, consensus and integrated developed so far.

Mapping of simply inherited traits and quantitative traits controlled by oligogenes and polygenes, respectively has been deliberated in the earlier literature crop-wise or crop group-wise. However, more detailed information on mapping or tagging oligogenes by linkage mapping or bulked segregant analysis, mapping polygenes by QTL analysis, and different computer software employed such as MapMaker, JoinMap, QTL Cartographer, Map Manager, etc. for these purposes have been discussed at more depth in the present volumes.

The strategies and achievements of marker-assisted or molecular breeding have been discussed in a few books and reviews earlier. However, those mostly deliberated on the general aspects with examples drawn mainly from major plants. In this series, we included comprehensive descriptions on the use of molecular markers for germplasm characterization, detection and maintenance of distinctiveness, uniformity and stability of genotypes, introgression and pyramiding of genes. We have also included elucidations on the strategies and achievements of transgenic breeding for developing genotypes particularly with resistance to herbicide, biotic and abiotic stresses; for biofuel production, biopharming, phytoremediation; and also for producing resources for functional genomics.

A number of desirable genes and QTLs have been cloned in plants since 1992 and 2000, respectively using different strategies, mainly positional cloning and transposon tagging. We included enumeration of these and other strategies for isolation of genes and QTLs, testing of their expression and their effective utilization in the relevant volumes.

Physical maps and integrated physical-genetic maps are now available in most of the leading crop and model plants owing mainly to the BAC, YAC, EST and cDNA libraries. Similar libraries and other required genomic resources have also been developed for the remaining crops. We have devoted a section on the library development and sequencing of these resources; detection, validation and utilization of gene-based molecular markers; and impact of new generation sequencing technologies on structural genomics.

As mentioned earlier, whole genome sequencing has been completed in one model plant (Arabidopsis) and seven economic plants (rice, poplar, peach, papaya, grapes, soybean and sorghum) and is progressing in an array of model and economic plants. Advent of massively parallel DNA sequencing using 454-pyrosequencing, Solexa Genome Analyzer, SOLiD system, Heliscope and SMRT have facilitated whole genome sequencing in many other plants more rapidly, cheaply and precisely. We have included extensive coverage on the level (national or international) of collaboration and the strategies and status of whole genome sequencing in plants for which sequencing efforts have been completed or are progressing currently. We have also included critical assessment of the impact of these genome initiatives in the respective volumes.

Comparative genome mapping based on molecular markers and map positions of genes and QTLs practiced during the last two decades of the last century provided answers to many basic questions related to evolution, origin and phylogenetic relationship of close plant taxa. Enrichment of genomic resources has reinforced the study of genome homology and synteny of genes among plants not only in the same family but also of taxonomically distant families. Comparative genomics is not only delivering answers to the questions of academic interest but also providing many candidate genes for plant genetic improvement.

The 'central dogma' enunciated in 1958 provided a simple picture of gene function—gene to mRNA to transcripts to proteins (enzymes) to metabolites. The enormous amount of information generated on characterization of transcripts, proteins and metabolites now have led to the emergence of individual disciplines including functional genomics, transcriptomics, proteomics and metabolomics. Although all of them ultimately strengthen the analysis and improvement of a genome, they deserve individual deliberations for each plant species. For example, microarrays, SAGE, MPSS for transcriptome analysis; and 2D gel electrophoresis, MALDI, NMR, MS for proteomics and metabolomics studies require elaboration. Besides transcriptome, proteome or metabolome QTL mapping and application of transcriptomics, proteomics and metabolomics in genomics-assisted breeding are frontier fields now. We included discussions on them in the relevant volumes.

The databases for storage, search and utilization on the genomes, genes, gene products and their sequences are growing enormously in each second and they require robust bioinformatics tools plant-wise and purpose-wise. We included a section on databases on the gene and genomes, gene expression, comparative genomes, molecular marker and genetic maps, protein and metabolomes, and their integration.

Notwithstanding the progress made so far, each crop or model plant species requires more pragmatic retrospect. For the model plants we need to answer how much they have been utilized to answer the basic questions of genetics and genomics as compared to other wild and domesticated species. For the economic plants we need to answer as to whether they have been genetically tailored perfectly for expanded geographical regions and current requirements for green fuel, plant-based bioproducts and for improvements of ecology and environment. These futuristic explanations have been addressed finally in the volumes.

We are aware of exclusions of some plants for which we have comprehensive compilations on genetics, genomics and breeding in hard copy or digital format and also some other plants which will have enough achievements to claim for individual book volume only in distant future. However, we feel satisfied that we could present comprehensive deliberations on genetics, genomics and breeding of 30 model and economic plants, and their groups in a few cases, in this series. I personally feel also happy that I could work with many internationally celebrated scientists who edited the book volumes on the leading plants and plant groups and included chapters authored by many scientists reputed globally for their contributions on the concerned plant or plant group.

We paid serious attention to reviewing, revising and updating of the manuscripts of all the chapters of this book series, but some technical and formatting mistakes will remain for sure. As the series editor, I take complete responsibility for all these mistakes and will look forward to the readers for corrections of these mistakes and also for their suggestions for further improvement of the volumes and the series so that future editions can serve better the purposes of the students, scientists, industries, and the society of this and future generations.

Science publishers, Inc. has been serving the requirements of science and society for a long time with publications of books devoted to advanced concepts, strategies, tools, methodologies and achievements of various science disciplines. Myself as the editor and also on behalf of the volume editors, chapter authors and the ultimate beneficiaries of the volumes take this opportunity to acknowledge the publisher for presenting these books that could be useful for teaching, research and extension of genetics, genomics and breeding.

Chittaranjan Kole

Preface to the Volume

The commercial berry has enjoyed a resurgence of interest over the last decade, mostly due to the promotion of its potentially positive health benefits. Recent discussions have implicated berries in the possible mitigation of aging, cancer, brain function, cardiovascular disease and metabolic disorders. Colorful and flavorful berries are capsules filled with antioxidants, anthocyanins, polyphenols and ellagic acid—all compounds with possible roles in disease prevention or enhanced quality of life. Berries occupy a distinct dietary role in highly industrialized nations. As mundane causes of mortality are overcome with new medical technology humans are living longer, and the trends predict higher likelihoods of long-term degenerative disease leading to death. Diets rich in fruits and vegetables will have a central role in limiting or preventing these illnesses, if not for their inherent compounds then for their intense flavors and colors with few accompanying calories. In the developing world berries might deliver nutritious compounds to those who desperately need them- possibly via plants that will be adapted to local conditions. The popularity and demand for berries continues to increase.

At the same time the production of all fruits, nuts and vegetables is limited by a series of challenges. Once potent soil fumigants have been banned, no longer presenting an affordable and effective way to curb soil pests, pathogens and weedy competitors. The price of labor, fuel, and fertilizers continues to rise, while prime land and fresh water become increasingly scarce. Farmers have to do more with less, and then do so in a market where prices are depressed by a limited number of competitive wholesalers. The current slate of commercial cultivars meets the berry demand, but as consumption grows in concert with production threats, new cultivars will need to be developed to meet the challenges of production and distribution.

Today, the challenge is to enhance varietal performance and produce perfect products with fewer agricultural inputs. These accomplishments must be performed with high sensitivity to environmental stewardship yet be flexible in an ever-changing climate. Traditional crop breeding offers the potential to greatly improve the capacity of cultivars. Coupled with modern genomics-enabled breeding techniques, the likelihood for breeding programs to deliver better products faster only increases. Plant breeding has always been as much as an art as it is science, and now the effective staples of genetics and selection find new partnership with high-throughput information capture and in planta validation.

The central challenge of the post-genomics era will be to marry copious genetic information to the breeding task, to sculpt it into productive channels that will drive the efficient production of higher-quality fruits. This book explores the state of the art of berry genomics, genetics and breeding and provides some hints as to how these technologies can assist, or in some cases have assisted, in berry improvement.

In order to best define the scope of this text it is important to examine some definitions and delineate what is and what is not a berry in the parlance of this work. A botanist will maintain strict adherence to a textbook definition where a berry is a simple fruit containing seeds produced from a single ovary. Some examples of true botanical berries are tomatoes, plantains, eggplants, grapes, persimmons, loquats and peppers. A strictly botanical analysis of the table of contents will reveal that this book is not really about berries at all, but instead is a text on epigynous, aggregate and accessory fruits. While botanically accurate, such descriptions create a cognitive disconnect from the spirit and intent of this work and hardly make for a compelling title. This text is about the berry of the familiar vernacular, the small, flavorful and nutritious plant-borne foods, rich in sugars and healthful compounds.

Another contradiction: Many true berries certainly do reside within the pervue of this strictly botanical definition. True berries, such as currants and gooseberries (*Ribes* sp.), elderberries (*Sambucus* sp.) or mulberries (*Morus* sp.) are small flavorful fruits, also ascending to similar superfood status. Recent advances in currant breeding have greatly increased the size of the berry and yield of this crop. Breeding of white and black mulberries is an important venture in many parts of the world, again with measurable gains and outstanding products. Unfortunately, these true berries are represented by only skim treatment with genetic, molecular or genomic tools. The mission is a text on the genetics and genomics of berries, but with no genomics coverage in most berry species, the work is forced to explore the growing data in the small fruits most commonly thought of as berries.

For the purposes of this text berries will be discussed in the most familiar sense, examining the advances in genetics and genomics of small, soft fruits. The subjects are a small set of species from the *Ericaceae* and *Rosaceae* that have enjoyed description and improvement through use of molecular and/or genomics tools. The work is weighted certainly to the species that have benefitted from the application of genomics tools. Strawberry (*Fragaria* spp.) has been an increasingly attractive system for genomics-level analyses in the last decade. There is relatively more information, so the chapter has been divided into three independent sections that each stand alone and benefit from encapsulation. Blueberries and cranberries are described in stand-alone chapters while *Rubus* species (blackberries and raspberries) are covered in a common chapter.

To conclude, this text on berries is paradoxically not about true berries at all, but instead is about a suite of valued plant products that are central to regional economies, human health and consumer choice. The genomics and post-genomics opportunities have only begun to manifest themselves in these species, and this text marks a stepping-off point for the inevitable gains soon to be realized. Outstanding traditionally-bred resources, a rich history of physiology and pathology, and a substantial body of work in production and post-harvest technology present many areas of genomics ingress, enabling another level of analyses to synergize with existing knowledge to accelerate the development of new germplasm. The confluence of technologies will inevitably lead to new plant products, but also new paradigms in biological processes yet to be discovered.

Kevin M. Folta

Contents

List of Contributors

Nahla V. Bassil
National Clonal Germplasm Repository,US Department of Agriculture, Agricultural Research Service, Corvallis, OR 97333, USA.

John E. Carlson
The School of Forest Resources, Department of Horticulture, and The Huck Institutes of Life Sciences, Pennsylvania State University,323 Forest Resources Building, University Park, PA 16802, USA.

Samir C. Debnath
Atlantic Cool Climate Crop Research Centre,Agriculture and Agri-Food Canada, P.O.Box 39088, 308 Brookfield Road, St. John's. Newfoundland and Labrador A1E 5Y7, Canada.

Béatrice Denoyes-Rothan
INRA UR 419, Unité de Recherche sur les Espèces Fruitières, Domaine de la Grande Ferrade, Villenave d'Ornon, France.

Felicidad Fernández-Fernández
East Malling Research,New Road, East Malling, Kent, ME19 6BJ, UK.

Chad E. Finn
US Dept of Agriculture- Agricultural Research, Service, Horticultural Crops Research Lab., 3420 NW Orchard Ave., Corvallis, OR 97330, USA.

Kevin M. Folta
Horticultural Sciences Department, University of Florida, Gainesville, FL, USA.

Julie Graham
SCRI, Invergowrie, Dundee, DD2 5DA, UK.

James F. Hanock
Department of Horticulture, Michigan State University, East Lansing, M1 48824, USA.

Amy Howell
Department of Plant Biology and Pathology, The School of Environmental and Biological Sciences, Rutgers, The State University of New Jersey, New Brunswick, NJ, USA.

Todd P. Michael
The Waksman Institute, Rutgers, The State University of New Jersey, Piscataway, NJ, USA.

Peter Oudemans
Department of Plant Biology and Pathology, The School of Environmental and Biological Sciences, Rutgers, The State University of New Jersey, New Brunswick, NJ, USA.

James Polashock
Genetic Improvement of Fruits and Vegetables Lab, USDA-ARS, Chatsworth, NJ, USA.

Mathieu Rousseau-Gueutin
INRA UR 419, Unité de Recherche sur les Espèces Fruitières, Domaine de la Grande Ferrade, Villenave d'Ornon, France.

Lisa J. Rowland
Genetic Improvement of Fruits and Vegetables Laboratory, Henry A. Wallace Beltsville Agricultural Research Center, US Department of Agriculture, Agricultural Research Service, Beltsville, MDS 20705, USA.

Daniel J. Sargent
East Malling Research, New Road, East Malling, Kent, ME19 6BJ, UK.

Janet P. Slovin
Genetic Improvement of Fruits and Vegetables Laboratory, Henry A. Wallace Beltsville Agricultural Research Center, US Department of Agriculture, Agricultural Research Service, Beltsville, MDS 20705, USA.

Philip J. Stewart
Driscoll's Strawberry Associates, Watsonville, CA, USA.

John-David Swanson
Department of Biology, University of Central Arkansas, 201 Donaghey Ave., Conway, AR 72032, USA.

Nicholi Vorsa
Department of Plant Biology and Pathology, The School of Environmental and Biological Sciences, Rutgers, The State University of New Jersey, New Brunswick, NJ, USA.

Courtney Weber
Department of Horticultural Sciences, Cornell University, New York State Agricultural Experiment Station, Geneva, NY 14456, USA.

Anna Zdepski
Department of Plant Biology and Pathology, The School of Environmental and Biological Sciences, Rutgers, The State University of New Jersey, New Brunswick, NJ, USA.

List of Abbreviations

4-coumarate:CoA ligase (4CL)
Amplified fragment length polymorphism (AFLP)
Bacterial artificial chromosome (BAC)
Blueberry Genomics Database (BBGD)
Blueberry shoestring virus (BBSSV)
Blueberry scorch virus (BlScV)
Blueberry shock virus (BlShV)
Black raspberry necrosis virus (BRNV)
Bulk-segregant-analysis (BSA)
Bacillus thurengensis (Bt)
Cold acclimated (CA)
Cleaved amplified polymorphic sequences (CAPS)
Chalcone synthase (CHS)
Endo-polygalacturonases (endo-PGs)
Expressed sequence tag (EST)
Expressed sequence tag-polymerase chain reaction (EST-PCR)
Flavanone 3-hydroxylase (F3H)
Genome Database for Rosaceae (GDR)
Green fluorescent protein (GFP)
Genetically modified organism (GMO)
Gene Pair Haplotype (GPH)
β-glucuronidase (GUS)
Hectares (ha)
High density lipoprotein (HDL)
High performance liquid chromatography (HPLC)
Hygromycin phosphotransferase (*hpt*)
I hate abbreviations (IHA)
Individually quick-frozen (IQF)
Inter-simple sequence repeat (ISSR)
Internal transcribed spacer of rDNA (ITS)
Leucine amino peptidase (LAP)
Linkage group (LG)
Marker assisted breeding (MAB)

Megabase (Mb)
Megabase pairs (Mbp)
Non-acclimated (NA)
National center for biotechnology information (NCBI)
Phenylalanine ammonia-lyase (PAL)
Phosphoglucoisomerase (PGI)
Polygalacturonase-inhibiting proteins (PGIPs)
Phosphoglucomutase (PGM)
Polyketide synthases (PKS)
Pectinmethyl esterase hydrolases (PME)
Phytophthora root rot (PRR)
Quantitative trait loci (QTL)
Randomly amplified polymorphic DNA (RAPD)
Raspberry bushy dwarf virus (RBDV)
Restriction fragment length polymorphism(RFLP)
Reverse subtracted library (RL)
Raspberry leaf mottle virus (RLMV)
Raspberry leaf spot virus (RLSV)
RNA interference (RNAi)
Raspberry vein chlorosis virus (RVCV)
Rubus yellow net virus (RYNV)
S-adenosylmethionine hydrolase (SAMase)
Sequence characterized amplified region (SCAR)
Scottish Crop Research Institute (SCRI)
Forward subtracted library (SL)
Single nucleotide polymorphism (SNP)
Single strand conformational polymorphism (SSCP)
Simple sequence length polymorphisms (SSLPs)
Simple sequence repeat (SSR)
Total anthocyanins (TAcy)
Tomato ringspot virus (TmRSV)
United States Department of Agriculture (USDA)
Variable number terminal repeat (VNTR)
Yeast artificial chromosome (YACs)

Blueberry

Lisa J. Rowland,[1]* *James F. Hancock,*[2] *and Nahla V. Bassil*[3]

1.1 Introduction

Blueberry is a high value crop which can thrive on acidic, imperfectly drained sandy soils, that might otherwise be considered worthless for agricultural crop production, and North America is the major producer of blueberries. Generally, cultivated blueberries belong to the section *Cyanococcus* of the genus *Vaccinium* of the heath family *Ericaceae* (Galletta and Ballington 1996). Species within this section are often called the "true" or cluster-fruited blueberries (Camp 1945). Wild representatives of *Cyanococcus* are found solely in North America (Hancock and Draper 1989). Blueberry species are also commonly grouped according to stature and referred to as the lowbush, highbush, and rabbiteye types. Lowbush plants are rhizomatous with stems from 0.30 to 0.60 m; highbush plants are crown forming and generally maintained between 1.8 and 2.5 m; rabbiteye plants are crown forming, but also are notable for suckering to varying degrees, and maintained between 2.0 and 4.0 m (Hancock and Draper 1989; Galletta and Ballington 1996). Of the major fruit crops, blueberry has been domesticated most recently, during the 20th century.

1.2 Basic Information on the Plant

1.2.1 Economic Importance

Several species of *Vaccinium* are important commercially. Most production comes from species in the section *Cyanococcus* including cultivars of

[1]Genetic Improvement of Fruits and Vegetables Laboratory, Henry A. Wallace Beltsville Agricultural Research Center, US Department of Agriculture, Agricultural Research Service, Beltsville, MD 20705, USA.
[2]Department of Horticulture, Michigan State University, East Lansing, MI 48824, USA.
[3]National Clonal Germplasm Repository, US Department of Agriculture, Agricultural Research Service, Corvallis, OR 97333, USA.
*Corresponding author: *Jeannine.Rowland@ars.usda.gov*

V. corymbosum L. (highbush blueberry) and *V. virgatum* Ait. (rabbiteye blueberry; syn. *V. ashei* Reade), and native stands of *V. angustifolium* Ait. (lowbush blueberry). Highbush cultivars are further separated into northern or southern types depending on their chilling requirements and winter hardiness.

All species of *Vaccinium* are acidophilic and woody perennials. Cultivated *Vaccinium* are asexually propagated through stem or rhizome cuttings or by micropropagation. All species of *Vaccinium* are invaded by mycorrhizal organisms.

Many of the wild, edible *Vaccinium* species have been harvested for thousands of years by indigenous peoples (Moerman 1998). Native Americans in western and eastern North America intentionally burned native stands of blueberries and huckleberries to renew their vigor. Highbush and rabbiteye blueberries were domesticated at the end of the 19th century. Plants were initially dug from the wild and transplanted into New England and Florida fields.

Most of the commercial production of blueberry comes from highbush and lowbush types, although rabbiteyes are important in the North American southeast and hybrids of highbush x lowbush (half-highs) have made a minor impact in the upper midwest of the USA. Rabbiteye cultivars are beginning to be grown in the Pacific Northwest and Chile for their very late ripening fruit. Highbush blueberries are grown in 37 states in the USA, in six Canadian provinces, and in Australia, Chile, Argentina, New Zealand and a number of countries in Europe (Strik 2005; Strik and Yarborough 2005). The largest acreages of northern highbush are in Michigan, New Jersey, North Carolina, Oregon, and Washington in the USA, and British Columbia in Canada. The greatest amount of southern highbush acreage is in Georgia, Florida, and California. Commercial production of lowbush blueberries is mainly in Maine, Quebec, New Brunswick, and Nova Scotia (Strik and Yarborough 2005). While the half-high blueberries are not a major contributor to the fruit market, they are very widely used as an ornamental plant for landscaping.

Over 90,000 t of highbush fruit are produced annually in the USA on over 20,000 ha (USDA Agricultural Statistics). The estimated area of rabbiteye production is currently about 3,000 ha, with half the acreage in Georgia. The total annual production is over 5,500 t. Half-high production is restricted to a few hundred hectares in Minnesota and Michigan. Annual production of lowbush blueberries ranges from 40,000 to 55,000 t on about 40,000 ha in primarily Maine and the Maritime provinces of eastern Canada.

1.2.2 Nutritional Composition

Blueberries are eaten as fresh fruit and in processed forms. About 50% of the highbush crop is marketed fresh and the remainder is processed. Individually quick-frozen (IQF), pureed, juiced, and dried/freeze-dried fruit are the primary processed products and from these a myriad of products appear in grocery stores.

An average blueberry fruit is composed of approximately 83% water, 0.7% protein, 0.5% fat, 1.5% fiber and 15.3% carbohydrate (Hancock et al. 2003). Blueberries have 3.5% cellulose and 0.7% soluble pectin, while cranberries contain 1.2% pectin. The total sugars of blueberries amount to more than 10% of the fresh weight, and the predominant reducing sugars in blueberries are glucose and fructose, which represent 2.4%. The primary organic acid in blueberries is citric acid (1.2%). Blueberries contain 22.1 mg of vitamin C per 100 g of fresh weight and cranberries contain 7.5–10.5 mg. Blueberries are unusual in that arginine is their most prominent amino acid.

In general, blueberries are one of the richest sources of antioxidant phytonutrients among fresh fruits, with total antioxidant capacity ranging from 13.9 to 45.9 μmol Trolox equivalents/g fresh berry (Ehlenfeldt and Prior 2001; Conner et al. 2002a, b). Total anthocyanins in blueberry fruit range from 85 to 270 mg per 100 g, and species in the subgenus *Cyanococcus* carry the same predominant anthocyanins, aglycones and aglycone-sugars, although the relative proportions vary (Ballington et al. 1988). The predominant anthocyanins are delphinidin-monogalactoside, cyanidin-monogalactoside, petunidin-monogalactoside, malvidin-monogalactoside, and malvidin-monoarabinoside. The major volatiles contributing to the characteristic aroma of blueberry fruit are *trans*-2-hexanol, *trans*-2-hexanal, and linalool (Hancock et al. 2003).

1.2.3 Academic Importance

Blueberry can serve as a model system for the study of autopolyploidy in plant species. Evolutionary biologists have typically recognized two general categories or types of polyploidy—allopolyploidy and autopolyploidy. Autopolyploids are generally considered to be those polyploids formed within or between populations of a biological species, whereas allopolyploidy involves those polyploids formed via hybridization between two distinct species. Allopolyploidy has long been considered prevalent in angiosperms, but the prevalence and importance of autopolyploidy in natural populations was traditionally questioned. Recent work has challenged this view and it is now recognized that autopolyploidy is a major evolutionary force and prominent type of speciation in plants (Soltis and Soltis 1993, 1999;

Soltis et al. 2007). As described below, *V. corymbosum* and its allied species, are a well documented polyploid complex that can be used as a platform to study the evolutionary dynamics of polyploidy.

In addition, blueberry has and continues to serve as a model system for studying cold tolerance in woody perennials. Winter injury is one of the most important factors limiting growth of woody perennials in North America (Quamme 1985). Periodic winter freezes result in serious losses in fruits, nuts, and ornamental crops. Consequently, a major emphasis of many breeding programs on woody plants is the development of more cold hardy cultivars. In a survey of all the blueberry research and extension scientists in the USA, lack of cold hardiness and susceptibility to spring frosts have been identified as the most important genetic limitations of current cultivars (Moore 1993). Genes for cold tolerance in blueberry, once identified and isolated, could be used for either blueberry transformation or marker-assisted selection to develop more cold hardy cultivars.

By far most of the genomic research on plant cold hardiness has been carried out on the herbaceous annual species, *Arabidopsis thaliana*. *Arabidopsis* does not undergo seasonal cold acclimation and acclimates only about 5–7°C, allowing for only brief exposures to freezing or near freezing temperatures (Wisniewski et al. 2003). Cold acclimation in woody perennials, on the other hand, is seasonal and is generally considered a two-step process, first triggered by shortening daylength and then declining temperatures (Weiser 1970; Powell 1987; Sakai and Larcher 1987). And many woody perennials can withstand extremely low subzero temperatures for extended periods of time. It is, therefore, likely that the mechanisms of cold acclimation in woody perennials are more complex than those in *Arabidopsis*, and warrant extensive investigation.

1.2.4 *Taxonomy and Germplasm Resources*

Overall, the genus *Vaccinium* is widespread, with high densities of species being found in the Himalayas, New Guinea and the Andean region of South America. The origin of the group is thought to be South American. Estimates of species numbers vary from 150–450 in 30 sections (Luby et al. 1991). Species delineation has been difficult to resolve in blueberries due to polyploidy, overlapping morphologies, continuous introgression through hybridization, and a general lack of chromosome differentiation. In the first detailed taxonomy of the *Cyanococcus* section or the "true" cluster-fruited blueberries, Camp (1945) described nine diploid, 12 tetraploid, and three hexaploid species, but Vander Kloet (1980, 1988) reduced this list to six diploid, five tetraploid and one hexaploid taxa. He included all the crown-forming species into *V. corymbosum* with three chromosome levels. Most horticulturists and blueberry breeders feel that the variation patterns in *V. corymbosum* are distinct enough to retain Camp's diploid *V. elliottii* Chapm. and *V. fuscatum* Ait., tetraploid *V. simulatum* Small, and hexaploid *V. constablaei* Gray and *V. virgatum* as separate species (Ballington 1990; Galletta and Ballington 1996; Ballington 2001; Lyrene 2008) For a listing of all the generally recognized *Vaccinium* species within the *Cyanococcus* section, see Table 1-1.

Table 1-1 Important species of blueberry, *Vaccinium* section *Cyanococcus* along with their ploidy levels and general distribution.

Species	Ploidy	Location
V. angustifolium Ait.	4*x*	N.E. North America
V. boreale Hall & Aald.	2*x*	N.E. North America
V. constablaei Gray	6*x*	Mountains of S.E. North America
V. corymbosum L.	2*x*	S.E. North America
V. corymbosum L.	4*x*	E. North America
V. darrowii Camp	2*x*	S.E. North America
V. elliottii Chapm.	2*x*	S.E. North America
V. fuscatum Ait.	2*x*	Florida
V. hirsutum Buckley	4*x*	S.E. North America
V. myrsinites Lam.	4*x*	S.E. North America
V. myrtilloides Michx.	2*x*	Central North America
V. pallidum Ait.	2*x*, 4*x*	Mid-Atlantic North America
V. simulatum Small	4*x*	S.E. North America
V. tenellum Ait.	2*x*	S.E. North America
V. virgatum Ait.	6*x*	S.E. North America

All the polyploid *Cyanococcus* are likely to be of multiple origin and active introgression between species is ongoing. The primary mode of speciation in *Vaccinium* has likely been through unreduced gametes, as there is a strong but not complete triploid block (Lyrene and Sherman 1983; Vorsa and Ballington 1991). The unreduced gametes are produced primarily through first division restitution (Qu and Hancock 1995; Qu and Vorsa 1999), although some second division restitution occurs as well (Vorsa and Rowland 1997). Embryo culture was not successful in recovering triploids of *V. elliottii* x tetraploid highbush (Munoz and Lyrene 1985), although triploids have been recovered through sexual reproduction (Vorsa and Ballington 1991).

The tetraploid highbush blueberry *V. corymbosum*, has been shown to be genetically an autopolyploid (Draper and Scott 1971; Krebs and Hancock 1989; Qu et al. 1998). There has apparently been little genomic evolution within the section *Cyanococcus*, as the tetraploid hybrids formed between diploid and tetraploid species are highly fertile (Draper et al. 1972), and a hybrid between evergreen, diploid *V. darrowii* Camp and deciduous, tetraploid *V. corymbosum* has been shown with randomly amplified polymorphic DNA (RAPD) markers to be undergoing regular, tetrasomic inheritance (Qu and Hancock 1995).

Interspecific hybridization within *Vaccinium* section *Cyanococcus* has played a major role in the development of highbush blueberries (Ballington 1990, 2001). Most homoploids freely hybridize and interploid crosses are frequently successful, through unreduced gametes (Lyrene et al. 2003). Genotypes have been found in many blueberry species that produce unreduced gametes (Ballington and Galletta 1976; Cockerman and Galletta 1976; Ortiz et al. 1992), and colchicine can be used to produce fertile genotypes with doubled chromosome numbers (Perry and Lyrene 1984). Even pentaploid hybrids of diploid x hexaploid crosses have been shown to cross relatively easily to tetraploids (Jelenkovic 1973; Chandler et al. 1985; Vorsa et al. 1987).

Numerous interspecific crosses have been made by breeders within section *Cyanococcus* including: 1) tetraploid *V. corymbosum* x tetraploid *V. angustifolium* (Luby et al. 1991), 2) tetraploid *V. myrsinites* Lam. x tetraploid *V. angustifolium* and *V. corymbosum* (Darrow 1960; Draper 1977), 3) colchicine-doubled diploid hybrids of *V. myrtilloides* Michx. x tetraploid *V. corymbosum* (Draper 1977), 4) diploid *V. darrowii* x hexaploid *V. virgatum* (Darrow et al. 1954; Sharp and Darrow 1959) and 5) diploid *V. elliottii* x tetraploid highbush cultivars (Lyrene and Sherman 1983). Probably the most widely employed interspecific hybrid has been US 75, a tetraploid derived from the cross of diploid *V. darrowii* selection Fla 4B x the tetraploid highbush cultivar Bluecrop. In spite of its being a hybrid of an evergreen, diploid species crossed with a deciduous, tetraploid highbush, US 75 is completely fertile and is the source of the low chilling requirement of many southern highbush cultivars (Draper and Hancock 2003). Some believe that Fla 4B may not be pure *V. darrowii*, but rather *V. darrowii* introgressed with 2x *V. corymbosum*.

Intersectional crosses have generally proved difficult, although partially fertile hybrids have been derived from *V. tenellum* Ait. and *V. darrowii* (section *Cyanococcus*) x *V. stamineum* L. (section *Polycodium*) (Lyrene and Ballington 1986), *V. darrowii* and *V. tenellum* x *V. vitis-idaea* L. (section *Vitis idaea*) (Vorsa 1997), *V. darrowii* x *V. ovatum* Pursh (section *Pyxothamnus*), *V. arboreum* Marshall (section *Batodendron*) and *V. stamineum* (section *Polycodium*) (Ballington 2001), and tetraploid *V. uliginosum* L. (section *Vaccinium*) x highbush cultivars (Rousi 1963; Hiirsalmi 1977; Czesnik 1985). Genes of *V. arboreum* have also been moved into tetraploid southern highbush using *V. darrowii* as a bridge (Lyrene 1991; Brooks and Lyrene 1998a, b). Genes from *V. ovatum* have been incorporated into ornamental highbush selections in the USDA-ARS Oregon program via NC 3048 (C. Finn, pers. comm.).

1.3 Classical Genetics and Traditional Breeding

1.3.1 Classical Breeding Achievements

Most of the blueberry breeding activity is focused on highbush blueberries, although a few programs are concerned with rabbiteye (Chile and Georgia) and half-high (Minnesota) types. Northern highbush blueberries are being bred in Australia, Chile and the USA (Arkansas, North Carolina, New Jersey, Michigan, Oregon, and Maryland). Southern highbush are being bred in Australia, Chile, and the USA (Florida, Mississippi, and North Carolina). The lowbush industry is based primarily on wild clones, and as a result there has been little breeding done on them, all of it in Nova Scotia.

The fruit characteristics most sought after in blueberries are flavor, large size, light blue color (a heavy coating of wax), a small scar where the pedicel detaches, easy fruit detachment for hand or machine harvest, firmness, and a long storage life. Other important characteristics are uniform shape, size, and color, high aroma, and ability to retain texture in storage. High levels of heritability have been identified for most of these traits and much genetic improvement has been made through conventional breeding (Luby et al. 1991; Galletta and Ballington 1996; Hancock et al. 2008). *V. darrowii* has been a particularly important source of powder blue color, intense flavor, and fruit that remains in good condition in hot weather (Ehlenfeldt et al. 1995; Ballington 2001; Draper and Hancock 2003). Figure 1-1 shows highbush blueberry fruit on the new cv. Draper, demonstrating

Figure 1-1 Highbush blueberry fruit on the new cv. Draper.
Color image of this figure appears in the color plate section at the end of the book.

the desirable characteristics of large size, light blue color, and relatively uniform ripening.

High antioxidant capacity has become an important fruit quality parameter in blueberries, although specific breeding for this characteristic has not yet been undertaken. Genetic improvement could be rapid, as considerable amounts of quantitative variability have been observed in this characteristic (Ehlenfeldt and Prior 2001; Connor et al. 2002a, b).

Most blueberry breeding programs are concerned with expanding the harvest season. Earliness is at a particular premium in the southern parts of the USA, Spain, Argentina and north-central Chile, while lateness is extremely important in Michigan and the Pacific Northwest. Increases in earliness have been successfully achieved by selecting for earlier bloom dates and shorter ripening periods, while lateness has been increased primarily by selecting individuals with very slow rates of fruit development. Bloom date, ripening interval, and harvest dates are highly heritable in blueberry populations (Lyrene 1985; Finn and Luby 1986; Hancock et al. 1991), with strong genotype by environmental interactions (Finn et al. 2003). Bloom date is strongly correlated with ripening date, but early ripening cultivars have been developed that have later than average flowering dates such as "Duke" and "Spartan" (Hancock et al. 1987).

The most desirable highbush and rabbiteye bush habit is one that is upright, open and vase shaped, with a height of 1.5 to 2.0 m and a modest number of renewal canes. Many cultivars have been developed that meet this ideotype. In general, plant height appears to be quantitatively inherited, although the short stature of *V. angustifolium* and *V. darrowii* is dominant to highbush in many interspecific crosses (Johnston 1946; Luby and Finn 1986; Lyrene 2008). In fact, high percentages of dwarf plants are found in many southern highbush breeding populations. Rabbiteye breeding populations are all upright and tall growing, with most being much taller than the highbush types.

Another important architectural feature in blueberries is an open flower cluster that is easily picked. Long pedicels and peduncles are the major components of this feature. While no formal genetic studies have been conducted on these traits, there appears to be considerable genetic variability in the primary gene pool being used by breeders, as many loose-clustered cultivars have been developed.

Expanding the range of adaptation of the northern highbush blueberry by reducing its chilling requirement has been an important breeding goal for over 50 years. This has been successfully accomplished by incorporating genes from the southern diploid species *V. darrowii* into *V. corymbosum* via unreduced gametes, although hybridizations with native southern *V. corymbosum* and *V. virgatum* have also played a role. Cultivars with almost a continuous range of chilling requirements are now available from 0–1,000 hours. The genetics of the chilling requirement has not been formally determined; however, segregation patterns suggest that it is largely quantitatively inherited with low chilling requirement showing additive and perhaps some dominance gene action (Rowland et al. 1998, 1999).

Winter cold often causes severe damage to blueberry flower buds and young shoots in the colder production regions. In general, northern highbush types survive much colder mid-winter temperatures than rabbiteye and southern highbush cultivars, although considerable variability exists within groups that have been exploited by breeders (Hancock et al. 1997; Ehlenfeldt et al. 2003, 2006, 2007; Rowland et al. 2005; Ehlenfeldt and Rowland 2006; Hanson et al. 2007). In full dormancy, northern highbush genotypes have been found to range in tolerance from –20 to –30°C, while rabbiteye genotypes range from –13 to –24°C (Ehlenfeldt et al. 2006). Few southern highbush have been evaluated, although "Legacy" has been found to tolerate temperatures to –17°C and "Ozarkblue" to –26°C (Rowland et al. 2005). "Sierra", which is composed of 50% southern germplasm has tolerated temperatures in excess of –32°C (Hancock, personal observation). The wood of half-high cultivars, such as "Northblue", can survive to –40°C and the flower buds can tolerate –36°C (C. Finn, pers. observ.).

In genetic studies on cold tolerance of blueberry (Arora et al. 1998, 2000; Rowland et al. 1999) it was found that the cold hardiness data in diploid populations fit a simple additive-dominance model of gene action, with the additive effects being greater than the dominance ones. Several wild species carry useful genes for cold hardiness including *V. angustifolium*, *V. boreale* Hall and Aald., *V. myrtilloides* (Galletta and Ballington 1996) and *V. constablaei* (Rowland et al. 2005; Ehlenfeldt and Rowland 2006; Ehlenfeldt et al. 2007). The cold-responsive proteins known as dehydrins also play a significant role in the cold hardiness of blueberries, as will be discussed in depth later (Muthalif and Rowland 1994; Arora et al. 1997; Levi et al. 1999; Panta et al. 2001; Rowland et al. 2004; Dhanaraj et al. 2005).

Spring frosts commonly damage flower buds of all blueberry species. Overall, southern highbush flower buds and developing flowers appear to be more cold-tolerant than rabbiteye flower buds (Lyrene 2008) and northern highbush flower buds tend to be more tolerant than southern highbush types. However, the stage of floral development when a frost occurs is much more important than relative bud hardiness (Hancock et al. 1987). Those cultivars with late bloom dates tend to suffer less frost damage than those flowering earlier because frosts are less common when those cultivars are blooming. As previously mentioned, breeders have produced a number of early ripening cultivars with later bloom dates that can avoid frost damage most years (e.g., "Duke" and "Spartan").

Rate of deacclimation may also play a role in early spring flower bud tolerance, and patterns of variability are great (Ehlenfeldt et al. 2003; Arora et al. 2004; Rowland et al. 2005). Rowland et al. (2005) found the northern highbush "Duke" to be the most rapid deacclimator of a mixed group of 12 cultivars, while the southern highbush "Magnolia", the northern highbush x rabbiteye pentaploid hybrid "Pearl River", the rabbiteye x *V. constablaei* cultivar "Little Giant", and the half-highs "Northcountry" and "Northsky" were the slowest. Northern highbush "Bluecrop" and "Weymouth", southern highbush "Legacy" and "Ozarkblue", and rabbiteye "Tifblue" were intermediate.

Flower buds can also be damaged in the northern production regions by rapid freezes in the fall. The flower buds of rabbiteye and southern highbush cultivars are generally considered to acclimate more slowly in the fall than those of northern highbush cultivars and as a result are more subject to late fall freezes (Hanson et al. 2007; Rowland et al. 2008b). Hanson et al. (2007) found that leaf retention in the fall was not a good predictor of rate of acclimation, as "Ozarkblue" and US 245 retain their leaves until very late in the fall, but are just as hardy as the mid-season standard "Bluecrop". Bittenbender and Howell (1975) also found no correlation between flower bud hardiness and fall leaf retention.

Another major focus of many blueberry breeders has been the combined adaptation to heat and drought. Most blueberry species are negatively impacted by high temperature and drought; however, rabbiteye types tolerate these conditions better than highbush, and southern highbush are generally superior to northern highbush (Galletta and Ballington 1996). Breeders have had some success in producing more heat tolerant cultivars, although the hottest temperatures of summer still have a major impact on the storage life of harvested fruit in all areas of blueberry production.

Among the other abiotic factors limiting blueberries, high pH and tolerance to mineral soils are very important. The *Vaccinium* are "acid-loving" and as such generally require soils below pH 5.8 for high vigor. Most blueberry breeders have not focused on this characteristic, although useful genetic variation likely exists in several wild species. Erb et al. (1990, 1993, 1994) discovered several interspecific hybrids that transmitted mineral soil adaptation including pentaploid JU-11 (*V. virgatum* x *V. corymbosum*), tetraploid JU-64 (tetraploid *V. myrsinites* x *V. angustifolium*) and tetraploid US 75 (*V. darrowii* x *V. corymbosum*). Scheerens et al. (1993a, b) also found that hybrids with JU-11, JU-64 and US 75 had mineral soil adaptation, along with "Jersey", "Sunrise" and complex hybrids of *V. elliottii*. Finn et al. (1993a, b) found progenies from *V. corymbosum*, *V. angustifolium* and *V. corymbosum* x *V. angustifolium* hybrids to vary in their pH tolerance, even though *V. angustifolium* was not generally a good source of tolerance.

Blueberries are routinely subject to a wide array of diseases (Caruso and Ramsdell 1995; Cline and Schilder 2006). Probably the most significant problems in highbush blueberry are mummy berry [*Monilinia vaccinii-*

corymbosi (Reade)], blueberry stunt phytoplasma, *Blueberry shoestring virus* (BBSSV), *Blueberry shock virus* (BlShV), *Tomato ringspot virus* (TmRSV), *Blueberry scorch virus* (BlScV), stem blight [*Botryosphaeria dothidea* (Moug.: Fr.) Ces and de Not.], stem or cane canker (*Botryosphaeria corticis* Demaree and Wilcox), *Phytophthora* root rot (*Phytophthora cinnamomi* Rands), Phomopsis canker (*Phomopsis vaccinii* Shear), *Botrytis* (*Botrytis cinerea* Pers.: Fr.), and anthracnose fruit rots (*Colletotrichum gloeosporioides* (Penz.) Penz. and Sacc.]. Most of these diseases are widespread, although mummy berry and the viral diseases are most prevalent in areas that grow northern highbush, and stem blight, cane canker, and *Phytophthora* root rot are most common in rainy, hot climates where southern highbush are grown. Fungal-induced defoliation is also a problem in the southeastern USA. Rabbiteye blueberries have somewhat different disease susceptibilities than highbush, but can be affected by *Botrytis* blossom and twig blight, stem blight and mummy berry, and several defoliating fungal diseases. Lowbush is most negatively impacted by *Botrytis* stem and twig blight, and red leaf disease caused by *Exobasidium vaccinii* (Fckl.) Wor.

Resistant or tolerant cultivars have been produced for most of the fungal diseases in highbush and rabbiteye blueberries, although the genetics of resistance has only been determined for *Phytophthora* root rot, *Phomopsis* canker, cane canker, and stem blight (Luby et al. 1991; Galletta and Ballington 1996). Inheritance of resistance to all these diseases is quantitative, with resistance to *Phytophthora* root rot being partially recessive (Clark et al. 1986). Resistance to stem blight and cane canker in Florida and North Carolina is so critical that high proportions of otherwise acceptable test clones are eliminated, because they have insufficient resistance (Lyrene 2008).

For viral diseases, only limited sources of resistance have generally been found. Cultivars show a range of responses to the Northwest strain of BlScV, whereas "Jersey" is the only cultivar that appears unaffected by the East Coast strain (Martin et al. 2006). Only field resistance has been found to BBSSV and it is limited to one cultivar, "Bluecrop" (Hancock et al. 1986). Several sources of resistance to *Blueberry red ringspot virus* were observed in highbush and rabbiteye blueberry (Ehlenfeldt et al. 1993).

A number of insects and arthropods cause significant damage to highbush blueberries including the blueberry maggot (*Rhagoletis pomonella* Walsh), blueberry gall midge (*Dasineura oxycoccana* Johnson), blueberry bud mite (*Acalitus vaccinii* Keifer), flower thrips (*Franklinellia* ssp.), Japanese beetle (*Popillia japonica* Newman), sharp-nosed leafhopper (stunt vector) *Scaphytopius magdalensis* Prov., blueberry aphid (BBSSV and BlScV vector) (*Illinoia pepperi* Mac. G.), cranberry fruit worm (*Acrobasis vaccinii* Riley), cherry fruit worm (*Grapholita packardi* Zell), the plum curculio (*Conotrachelus nenuphar* Herbst) and found for the first time in North America in 2008, the spotted wing Drosophila (*Drosophila suzukii* Matsumura). Flower thrips, blueberry bud mite, and the gall midge are particular problems in the southeastern USA. Lowbush and rabbiteye blueberries generally suffer from fewer major pests than highbush types; however, significant damage is caused by cranberry fruitworm and stunt in rabbiteye blueberry, and by maggots in lowbush blueberry.

Little variation in resistance has been reported to most of these pests in *Vaccinium*, except for sharp-nosed leafhopper, blueberry aphid, bud mite and gall midge. Most southern highbush cultivars have medium to high resistance to the blueberry gall midge and numerous cultivars exist that are resistant to the blueberry bud mite (Lyrene 2008). A wide range of densities of blueberry aphids were found on northern highbush cultivars, but no immunity was identified (Hancock et al. 1982). Ranger et al. (2006, 2007) found several *Vaccinium* species that were more resistant to aphid colonization and population growth than *V. corymbosum*. Resistance to the vector of blueberry stunt, the sharp-nosed leafhopper, has been found in *V. virgatum* and *V. elliottii*, but not in wild or cultivated *V. corymbosum*. The resistance to the sharp-nosed leafhopper is quantitatively inherited in *V. virgatum*, but is monogenic in *V. elliottii* (Meyer and Ballington 1990; Ballington et al. 1993).

1.3.2 Limitations of Traditional Breeding

Blueberry breeders have liberally employed interspecific hybridization in the development of both northern and southern highbush types. In these efforts, species have been blended with very different chilling requirements and cold tolerances. To evaluate the climatic adaptations of the hybrids in the field has proven quite difficult, as chilling hours and temperatures vary greatly from year to year and site to site, necessitating long trialing periods to adequately evaluate a genotype's adaptive zone. It has proven particularly difficult to separate those genotypes with extremely low chilling requirements from those with intermediate requirements, and to identify hybrids that acclimate and deacclimate to cold at rates appropriate for northern production regions. The identification of QTL regulating chilling hour requirements and cold tolerance would allow for the more precise prediction of a hybrid's adaptive range and lead to more rapid genetic improvement through marker-assisted breeding. Association analyses, combined with candidate gene approaches, would also aid in the identification of useful alleles in diverse hybrid populations.

Map-based approaches could also facilitate the breeding of early-ripening cultivars that are less subject to frost. The association between early blooming and ripening has proven problematic to breeders who are trying to develop early ripening types, because early bloomers are often damaged by spring frosts. The rate of floral development is likely a complex interaction between chilling requirements, rates of deacclimation from cold, and temperature thresholds for active growth. The mapping of relevant genes would allow for the identification of those genotypes with late bloom and early ripening characteristics that are in tune with local climatic conditions.

The identification and mapping of genes associated with fruit quality would also be highly desirable. Breeders seek cultivars with light blue, firm fruit that have tiny pedicel scars, excellent flavor, and a long storage life. Conventional breeding approaches have yielded many cultivars with fruit that are nicely colored, firm, high flavored with tiny scars, but the combination of excellent flavor and long storage life has proven more difficult to achieve. The longest stored cultivars often have very acidic fruit. Breeding progress in this regard could be greatly streamlined through marker-assisted breeding, if the key genes associated with flavor and storage life could be identified and mapped. Marker-assisted breeding could also aid highbush breeders in combining the genes associated with high fruit quality with those determining regionally appropriate chilling hour and cold requirements.

It would also be beneficial to have markers for the genes conferring resistance to the most common blueberry diseases. At present, although some controlled screenings for disease resistance have been done (Stretch et al. 1995; Ehlenfeldt and Stretch 2000; Stretch and Ehlenfeldt 2000; Polashock et al. 2005; Polashock and Kramer 2006), most resistance screening is done in the field utilizing natural field infestations. Disease pressure varies greatly across years and sites, making precise ratings of resistance difficult. Having tags to the most important resistance alleles would aid greatly in the identification of elite, resistant germplasm and make it easier to pyramid multiple resistances into single cultivars.

1.4 Diversity Analysis

RAPD markers have been used for DNA-fingerprinting representative selections and cultivars of the three major commercially grown types of blueberries, i.e., the highbush, lowbush, and rabbiteye types. Expressed sequence tag-polymerase chain reaction (EST-PCR) markers and cleaved amplified polymorphic sequences (CAPS) markers derived from EST-PCR markers have been used for DNA-fingerprinting representative selections and cultivars of mainly highbush blueberry types but also including a couple of rabbiteye, one lowbush *V. darrowii*, and several *V. angustifolium* genotypes. Blueberry microsatellite or simple sequence repeat (SSR) markers have been used for fingerprinting blueberry and cranberry cultivars. In addition, RAPD, EST-PCR and CAPS, SSRs, and isozyme markers have been used to assess the genetic relationships of various blueberry populations, selections and cultivars.

Hill and Vander Kloet (1983), in their initial efforts to identify isozyme markers for genetic studies in blueberry, reported limited variation in four enzyme systems (esterase, malate dehydrogenase, peroxidase and phosphoglucose isomerase) among four *Vaccinium* sections including *Cyanococcus*. The authors believed their difficulties in isozyme analysis in blueberry were due to phenolic interference. After refinement of techniques to reduce or eliminate this phenolic interference, several researchers reported successful recovery of blueberry isozymes by starch gel electrophoresis.

Bruederle et al. (1991) extended isozyme analyses of 20 loci to the investigation of population genetic structure among diploid blueberry species, *V. elliottii*, *V. myrtilloides* and *V. tenellum*. They found that the diploid species exhibit high levels of variation within populations as expected for highly self-sterile, outcrossing taxa. All populations were in Hardy-Weinberg equilibrium with slight heterozygote excess observed in the more broadly distributed *V. tenellum* and *V. myrtilloides*.

Bruederle and Vorsa (1994) employed isozyme data, collected at 11 polymorphic loci, to assess the genetic relationships of representative diploid blueberry populations. Genetic similarity values were calculated and a cluster analysis was performed on the similarity data matrix. The data supported the recognition of two highbush diploid taxa, *V. corymbosum* and *V. elliottii*, instead of only *V. corymbosum* as was proposed by Vander Kloet (1988). The lowbush species *V. boreale* and *V. myrtilloides* were not readily distinguishable from each other suggesting they diverged fairly recently.

Hokanson and Hancock (1998) examined levels of allozymic diversity in native Michigan populations of diploid *V. myrtilloides* and the tetraploids, *V. angustifolium* and *V. corymbosum*. Six enzyme systems were evaluated and levels of heterozygosity and the number of alleles were averaged over seven polymorphic isozyme loci. The level of heterozygosity and number of alleles per locus were significantly lower in the diploid *V. myrtilloides* than in the tetraploids. This is similar to what has been found in other studies comparing closely related diploid and tetraploid species.

Most commercially important improved rabbiteye (*V. virgatum*) cultivars were developed from only four original native selections from the wild (Aruna et al. 1993). Because of this very narrow germplasm base, Aruna et al. (1993) used RAPD markers generated from amplification with 20 RAPD primers to investigate the extent of genetic relatedness among 19 cultivars of rabbiteye blueberry, 15 improved cultivars and the four original selections from the wild. As expected, results showed that all the improved cultivars are progressing towards increased genetic similarity when compared with the initial four wild selections. In addition, clustering of genotypes based on genetic similarity estimates generally agreed with known pedigree information, grouping siblings with each other and with one or both parents. In an extension of this original study, Aruna et al. (1995)

developed a cultivar key for distinguishing the 19 rabbiteye cultivars based on 11 RAPD markers amplified from four RAPD primers.

Levi and Rowland (1997) used RAPD and inter-simple sequence repeat (ISSR) markers to differentiate and evaluate genetic relationships among 15 highbush (*V. corymbosum*) or highbush hybrid cultivars, two rabbiteye (*V. virgatum*) cultivars, and one southern lowbush (*V. darrowii*) selection from the wild. Fifteen RAPD and three ISSR markers were chosen to construct a DNA fingerprinting table to distinguish among the genotypes in the study. A cluster analysis, based on similarity coefficients calculated from the molecular marker data, effectively separated out the different species examined. However, clustering of genotypes within the *V. corymbosum* group did not agree well with known pedigree data. The authors cautioned against using RAPD or ISSR marker data alone to assess genetic relationships of cultivars or selections within a species. Arce-Johnson et al. (2002) reported using two RAPD primers to distinguish five highbush cultivars in Chile.

Burgher et al. (1998, 2002) screened 26 wild lowbush (*V. angustifolium*) clones, including six named cultivars and 12 selections, with 30 RAPD primers. All could be differentiated using 11 of the primers. Clustering of genotypes correlated fairly well with geographic origin of the clones.

More recently, Rowland et al. (2003c) used EST-PCR and EST-PCR-derived CAPS markers to differentiate and evaluate genetic relationships among 15 highbush (*V. corymbosum*) or highbush hybrid cultivars, two rabbiteye (*V. virgatum*) cultivars, and two wild selections (one *V. darrowii* and one diploid *V. corymbosum* selection), which are the original parents of a mapping population. A subset of four EST-PCR primer pairs were identified that were sufficient to distinguish all the genotypes. A fairly good correlation between the similarity coefficients calculated from molecular marker data and coefficients of coancestry calculated from pedigree information was found based on the rather small number of markers that were analyzed. Currently the highbush (*V. corymbosum*)-derived EST-PCR markers are being used in genetic diversity studies on wild lowbush (*V. angustifolium*) blueberry (Bell et al. 2008).

Blueberry SSR markers were used to identify blueberry (Boches et al. 2006a) and cranberry, *Vaccinium macrocarpon* Ait., cultivars (Bassil et al. 2008). Blueberry SSR markers were able to distinguish each of the 69 unique blueberry accessions and 16 important cranberry cultivars and group them according to pedigree. Microsatellite analysis reflected a high level of heterozygosity in the 69 blueberry accessions as indicated by an average of 17.7 alleles per single locus and a Shannon's index (H) of 9.77. SSRs were also used to assess genetic diversity in wild and cultivated highbush blueberry (Boches et al. 2006a). A statistically significant decrease in genetic diversity was observed among cultivated blueberries compared to wild blueberries. However, substantial genetic diversity was found in the cultivated blueberry gene pool. A similar conclusion was also reported by Brevis et al. (2008), where 21 single-locus SSRs were used to evaluate genetic relationships of southern highbush blueberry cultivars and to assess the effects of wide hybridization on the genetic diversity of these cultivars. Pedigree-based genetic distances and the SSR-based distance estimator, the proportion of shared alleles, were significantly correlated ($r = 0.57$, $P \leq 0.0001$), indicating that microsatellite markers are a reliable tool to assess the genetic relationships among southern highbush cultivars. It appears that strong selection pressure targeted on many loci has limited the introgression of rare alleles from non-cultivated *Vaccinium* species into southern highbush cultivars as indicated by similarity in two parameters, the molecular coancestry between southern highbush and historical northern highbush blueberry cultivars, and in the levels of heterozygosity between modern northern highbush and southern highbush cultivars. The relative genetic contributions of *V. angustifolium*, *V. corymbosum*, *V. darrowii*, *V. elliottii*, *V. tenellum*, and *V. virgatum* clones to 38 southern highbush cultivars were determined.

Phylogenetic relationships of 93 species, representing 28 genera and 16 sections within the blueberry tribe (Vaccinieae), have been analyzed based on sequence data from the chloroplast matK gene and the nrITS region (Kron et al. 2002). The study identified several well-supported clades within the tribe that do not, however, correspond to currently recognized taxonomic groups. *Vaccinium* was clearly polyphyletic, fragmenting throughout the tree. Only one representative from the section Cyanococcus was studied, *V. tenellum*.

1.5 Association Studies

To date, no attempts have been made to calculate the extent of linkage disequilibrium in blueberry, but as the mapping work progresses, these estimates should be forthcoming. Association mapping could prove to be extremely beneficial to blueberry breeding, as numerous complex hybrid populations have been generated that are likely a "gold mine" for unique alleles of horticultural importance.

1.6 Molecular Linkage Maps: Strategies, Resources, and Achievements

Development of genetic linkage maps and mapping of both simply inherited and quantitative traits in blueberry have lagged behind herbaceous annuals and some woody perennials for a number of reasons, including long generation times, high ploidy levels of commercial types, lack of described Mendelian markers, self- and cross-incompatibility, inbreeding depression, and recalcitrance to many molecular genetic and biochemical techniques (Rowland and Levi 1994; Rowland and Hammerschlag 2005). However, despite the difficulties in working with blueberry, much progress has been made in the last few years in developing more robust molecular markers for DNA-fingerprinting cultivars and selections, analysis of genetic diversity in breeding materials and wild populations, and constructing genetic linkage maps at the diploid and tetraploid levels.

1.7 Evolution of DNA Markers

Before the development of molecular markers, genetic analysis in blueberry, as in many perennial, outcrossing plant species, was severely constrained by the limited number of simply inherited genetic markers. Only four simply inherited traits had been described before DNA markers became available for blueberry (Lyrene 1988), for: glaucous leaf (dominant) versus nonglaucous in *V. angustifolium* (Aalders and Hall 1963); blue or black berry (dominant) versus albino berry in *V. myrtilloides* (Aalders and Hall 1962), *V. angustifolium* (Hall and Aalders 1963) and *V. corymbosum* (Ballinger et al. 1972); green seedling (dominant) versus lethal albino seedling in *V. corymbosum* (Draper and Scott 1971); and normally pigmented red fall foliage, reddish-brown bud scales, and black ripe fruit (dominant) versus anthocyanin-deficient yellow fall foliage, whitish-green bud scales, and greenish-white ripe fruit in *V. elliottii* (Lyrene 1988).

With the development of DNA marker techonologies, many methods have become available for generating large numbers of genetic markers. Initially, detection of RFLPs from chloroplast and mitochondrial DNA was tried on blueberry. Haghighi and Hancock (1992) analyzed restriction fragment length polymorphisms (RFLPs) in various genotypes representing the blueberry species, *V. angustifolium*, *V. virgatum*, *V. corymbosum*, and *V. darrowii* using chloroplast-specific and mitochondrial-specific gene probes. No polymorphisms were detected in the chloroplast genome, whereas high levels of polymorphism were observed in the mitochondrial genome. However, because of the disadvantages of RFLPs, namely the requirements for large amounts of DNA and Southern hybridizations of the digested DNA with radioactively-labeled probes, RFLPs never became a popular method for detecting polymorphisms in blueberry.

Soon after the PCR-based method for detecting DNA polymorphisms known as RAPD was developed (Welsh and McClelland 1990; Williams et al. 1990), RAPD markers became the marker-of-choice for blueberry. RAPD technology employs short 10-base primers of arbitrary nucleotide sequences (> 50% GC) for amplification of multiple fragments of genomic DNA that are easily visualized on ethidium bromide-stained agarose gels. There were several reasons for their initial popularity. No prior knowledge of the genome, such as sequence data or available cloned DNA, is required; and, at that time, these were not available for blueberry. Also, the procedure does not require the use of radioactive probes or as much DNA as RFLP analysis requires.

Aruna et al. (1993) and Levi et al. (1993) were the first to report successful amplification of RAPD markers from blueberry DNA. Aruna et al. (1993) reported good results from DNA of native selections and improved cultivars of rabbiteye blueberry (*V. virgatum*) using essentially the same amplification conditions described by Williams et al. (1990). Levi et al. (1993) described an optimized RAPD protocol for the reproducible amplification of RAPD markers from several different woody plants including blueberry, cherry, peach, pear, and apple. This procedure utilizes a PCR buffer that contains higher levels of gelatin (0.1%) and a nonionic detergent (1% Triton-X-100) than typically described, together with a higher annealing temperature of 48°C. Since this initial protocol was described, researchers have recommended replacing the gelatin in the buffer with 0.1% bovine serum albumin, after finding that the gelatin source could affect results (Stommel et al. 1997).

Despite their ease of use, however, RAPD markers have disadvantages over some other types of markers. Most RAPD markers are dominant, thus, less informative in some types of genetic analyses than codominant markers like isozymes and RFLP markers. Beyond that, RAPD markers have become notorious for being difficult to reproduce between laboratories. This has led many researchers to look for more robust marker systems to use.

SSRs or microsatellites (Weber and May 1989) are another type of PCR-based molecular marker, which utilize a subclass of repetitive DNA sequences containing iterations of very short simple sequence repeats (1–5 bp). Microsatellites are highly abundant in plant genomes and their loci are polymorphic (Wang et al. 1994). Once microsatellite loci are identified and sequenced, a pair of 20 base-long primers can be synthesized based on specific sequences flanking the microsatellites and used to amplify simple sequence length polymorphisms (SSLPs). They are robust, versatile markers that typically can be treated as codominant markers in diploids, but are often scored as dominant markers (presence or absence) in polyploids. In addition, they can be used in high-throughput situations combined with capillary electrophoresis. They have the limitation, however, that they are often species-specific or usable only in closely related species (Jarne and Lagoda 1996).

SSRs were developed for blueberry from a genomic library and two expressed sequence tag (EST) libraries derived from *V. corymbosum* cv. Bluecrop (Boches et al. 2005). Diversity parameters including average number of alleles, unique alleles, genotypes, and Shannon's index were slightly higher in genomic SSRs as compared to EST-SSRs but the difference was not statistically significant (Boches et al. 2006a). Thirty-six EST-SSRs were tested for cross-amplification in 23 genotypes from 10 sections of the genus *Vaccinium* (Boches et al. 2006b). Cross-amplification ranged from 17% to 100% and was 83% on average. EST-SSRs from *V. corymbosum* were most easily transferable to other members of the *Cyanococcus* section and least easily transferable to sections *Oxycoccus, Herpothamnus, Myrtillus,* and *Batodendron* in descending order. Several loci amplified exclusively in sections *Cyanococcus, Batodendron, Bracteata,* and *Ciliata,* indicating a possible genetic link between *Vaccinium* in the southeastern USA and those present in Asia. These blueberry SSRs were also evaluated in cranberry, *V. macrocarpon,* and 16 of the SSRs easily differentiated between 16 economically important cranberry cultivars and grouped them based on pedigree (Bassil et al. 2008). Since then, the blueberry genomic and EST-SSRs were evaluated for amplification and polymorphism in both the large-fruited cranberry (*V. macrocarpon*) and the small-fruited cranberry (*V. oxycoccos*) and the majority resulted in amplification (see Fig. 1.2; N. Bassil, unpubl). A larger proportion of EST-SSRs as compared to genomic SSRs amplified in both cranberry types. Of the EST-SSRs that amplified in cranberry, however, only 30–35% were not polymorphic while none of the genomic SSRs that amplified in *V. oxycoccus* and only 10% of those that amplified in *V. macrocarpon* lacked polymorphism. Higher cross-species transference of EST-SSRs as well as lower polymorphism as compared to genomic SSRs have been observed in other plants as well (Varshney et al. 2005). Recently, additional EST-SSRs have been developed from a private EST database of 9,011 unigenes developed by HortResearch (Wiedow et al. 2007). They have been used to assess genetic relationships among a small set of cultivated blueberries.

Other PCR markers derived from ESTs, termed EST-PCR markers, have also been developed for blueberry. These have a number of advantages for genetic studies. First, they target expressed genes; thus, they should be particularly useful for quantitative trait loci (QTL) mapping. If an EST marker is linked to a QTL, it is possible that the gene itself, from which the EST marker was derived, controls the trait in question. Second, because they are derived from gene coding regions, which are more likely to be conserved across populations and species than noncoding regions, EST markers should be useful for comparative mapping and phylogenetic analyses. Furthermore, EST-based markers have the potential for being codominantly inherited although, in reality, they are often scored as dominant markers when used with polyploids and/or when they produce multiple bands, where allelism cannot be easily determined.

In most plants, where EST-PCR markers have been tested, amplification using EST-specific primers must be followed by either digestion with restriction enzymes to generate CAPS markers, heteroduplex analysis, or

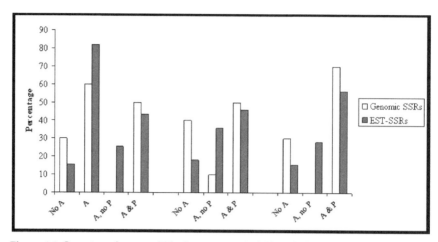

Figure 1-2 Cross-transference of blueberry genomic (10) and EST-SSRs (39) isolated from *V. corymbosum* cv. Bluecrop into seven accessions each of the large-fruited cranberry, *V. macrocarpon*, and the small-fruited cranberry, *V. oxycoccos*. Amplification in both species (A), as well as lack of amplification (No A), amplification but no polymorphism (A, no P) and amplification and polymorphism (A & P) in seven accessions of each species are estimated in both species and in each of *V. macrocarpon* and *V. oxycoccos* separately.

single-stranded conformational polymorphism (SSCP) analysis to detect polymorphisms. This is not the case, however, with blueberry, which is primarily outcrossing and highly heterozygous. Rowland et al. (2003c) designed 30 PCR primer pairs from ESTs from the highbush (*V. corymbosum*) cv. Bluecrop and tested them in amplification reactions with genomic DNA from 19 blueberry genotypes, including 15 highbush (*V. corymbosum*) or highbush hybrid cultivars, two rabbiteye (*V. virgatum*) cultivars, and two wild selections (one *V. darrowii* and one diploid *V. corymbosum* selection). Primers were designed near the ends of the ESTs to amplify as much of each gene as possible, to increase chances of detecting polymorphisms. Fifteen of the 30 primer pairs resulted in amplification of polymorphic fragments that were detectable directly after ethidium bromide staining of agarose gels. Several of the monomorphic amplification products were digested with the restriction enzyme *Alu*I and approximately half of these products resulted in polymorphic-sized fragments or CAPS markers. Since being developed for highbush (*V. corymbosum*) blueberry, the EST-PCR markers have been tested and shown to amplify fragments and detect polymorphisms in all the species within the section *Cyanococcus* (see Fig. 1.3; L. Rowland, unpubl). EST-PCR primer pairs have also been tested for their ability to amplify fragments in further related members of the *Ericaceae*, such as cranberry and rhododendron (Rowland et al. 2003a). Of the primer pairs tested in cranberry, 89% resulted in successful amplification and 35% of those amplified polymorphic fragments among the cranberry genotypes.

Of the primer pairs tested in rhododendron, 74% resulted in successful amplification and 72% of those amplified polymorphic fragments among the rhododendron genotypes. Thus, these markers should be useful for DNA fingerprinting, mapping, and assessing genetic diversity within cranberry and rhododendron species, in addition to blueberry species.

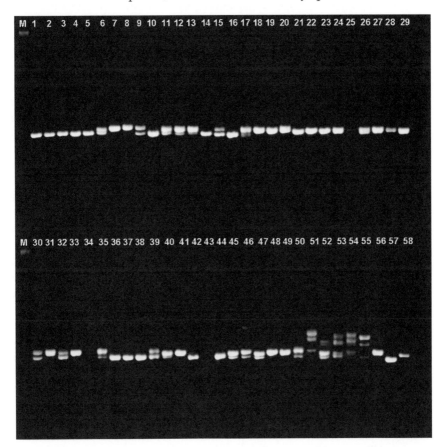

Figure 1-3 Agarose gel showing amplification products from EST-PCR primers designed from highbush blueberry-derived (*V. corymbosum*) EST CA661 (which encodes a dehydrin) tested on all *Vaccinium* species within the section *Cyanococcus*. The DNA samples were loaded as follows: 1-kb molecular weight ladder from Invitrogen Life Technologies (lane M), three *V. angustifolium* selections (lanes 1–3), three *V. boreale* selections (lanes 4–6), four *V. caesariense* Mack. selections (lanes 7–10), three *V. constablaei* selections (lanes 11–13), four tetraploid *V. corymbosum* selections (lanes 14–17), five *V. darrowii* selections (lanes 18–22), seven *V. elliottii* selections (lanes 23–29), three *V. hirsutum* selections (lanes 30–32), three *V. myrsinites* selections (lanes 33–35), three *V. myrtilloides* selections (lanes 36–38), two tetraploid *V. pallidum* selections (lanes 39–40), three diploid *V. pallidum* selections (lanes 41–43), three *V. simulatum* selections (lanes 44–46), three *V. tenellum* selections (lanes 47–49), six *V. virgatum* selections (lanes 50–55), two *V. fuscatum* Ait (syn *V. atrococcum* Gary [Heller]) selections (lanes 56–57), and one *Gaylussacia brachycera* (Michx) A. Gary selection to be used as an outlier (lane 58).

1.8 Construction of Genetic Linkage Maps

Efforts are underway to develop genetic linkage maps for blueberry that are saturated enough to map QTL controlling chilling requirement, cold hardiness, heat tolerance, and fruit quality. Relatively low density maps currently exist for five blueberry populations, which include three diploid and two tetraploid populations. Rowland and Levi (1994) reported construction of the first RAPD-based genetic linkage map of blueberry using a diploid population segregating for chilling requirement, which resulted from a true testcross between an F_1 interspecific hybrid (*V. darrowii* x *V. elliottii*) and another *V. darrowii* clone. A testcross was used because diploid blueberry species are essentially self-sterile and tolerate little inbreeding; therefore, true F_2 or backcrosses cannot be easily generated for mapping. The map was comprised of 72 RAPD markers that mapped to 12 linkage groups, in agreement with the basic chromosome number for blueberry.

Since that initial map was developed, Rowland et al. (1999) focused on construction of RAPD-based genetic linkage maps using diploid blueberry populations shown to segregate for both chilling requirement and cold hardiness. The populations resulted from true testcrosses between F_1 interspecific hybrids, *V. darrowii* x diploid *V. corymbosum*, and another *V. darrowii* clone and another diploid *V. corymbosum* clone. With the development of EST-PCR markers, however, focus has shifted to adding these types of markers to the maps. The maps of each of the *V. corymbosum* and *V. darrowii* testcross populations are currently comprised of approximately 100 RAPD and EST-PCR markers (Rowland et al. 2003b; L. Rowland unpubl).

Qu and Hancock (1997) reported construction of a RAPD-based genetic linkage map of a tetraploid blueberry population segregating for high fruit quality, heat tolerance, and cold tolerance. The population resulted from a cross of US 75 (a tetraploid hybrid of a diploid *V. darrowii* selection Fla 4B and tetraploid *V. corymbosum* "Bluecrop") and another *V. corymbosum* "Bluetta". One hundred and forty RAPD markers unique to Fla 4B that segregated 1:1 in the tetraploid population were mapped into 29 linkage groups. The map is essentially of *V. darrowii* because US 75 was produced via a 2*n* gamete from Fla 4B and only unique markers for Fla 4B were used (Qu and Hancock 1997). Interestingly, it is the same *V. darrowii* selection (Fla 4B) that was used as the original parent plant of the diploid mapping populations described earlier (Rowland and Levi 1994; Rowland et al. 1999). Fla 4B and its hybrids, such as US 75, have been used extensively in blueberry breeding programs to develop low-chilling southern highbush cultivars.

Most recently, Brevis et al. (2007) have been using the SSR markers of Boches et al. (2005 and 2006a, b) and EST-PCR markers of Rowland et al.

(2003c) to develop a linkage map of the tetraploid cross "Draper" (northern highbush) x "Jewel" (southern highbush). The ultimate goal is to identify QTL for chilling requirement, cold hardiness, and fruit quality which are segregating in this population.

1.9 Molecular Mapping of Simply Inherited and Complex Traits

Bulk-segregant-analysis (BSA) is currently being used to identify markers linked to such traits as parthenocarpy (L Rowland and M Ehlenfeldt unpubl) and mummy berry resistance (Polashock and Vorsa 2006) in blueberry. Parthenocarpy is the production of fruit without pollination. Parthenocarpy is advantageous in some crops when seeds are not desired by consumers or when pollination problems can arise. In most genotypes, blueberry seeds are not noticeable to consumers. However, for fruit development, bee pollination is required. In some years, weather conditions do not cooperate and conditions are excessively windy or rainy during the pollination window or, for other reasons, bee populations are low, resulting in major losses in fruit yield. A parthenocarpic mutant of highbush blueberry was selected several years ago in the USDA-ARS breeding program. Segregation ratios in crosses suggest that the parthenocarpic trait is controlled by a single recessive gene (Ehlenfeldt and Vorsa 2007). Polashock and Vorsa (2006) are using BSA to tag genes for mummy berry resistance in segregating blueberry populations with *V. darrowii* as the source of resistance.

Mapping efforts are also underway to identify QTL controlling complex traits such as chilling requirement, cold hardiness, heat tolerance, and fruit quality in blueberry. The *V. darrowii* x *V. corymbosum* interspecific testcross populations developed by Rowland et al. (1999) have been evaluated for chilling requirement and cold hardiness and a generation means analysis has been used to study the inheritance of cold hardiness in the cold-acclimated state (Arora et al. 2000). Results from the generation means analysis indicated that the cold hardiness data best fit a simple additive-dominance model of gene action (a model in which the genes controlling cold tolerance are assumed to have simple additive and dominance effects). Furthermore, in this study, the magnitude of the additive gene effect was greater than that of the dominance gene effect. A preliminary QTL analysis using the current genetic linkage map and cold hardiness data for the *V. corymbosum* testcross population has identified one putative QTL associated with cold hardiness that explains ~ 20% of the genotypic variance (Rowland et al. 2003b).

1.10 Molecular and Transgenic Breeding

Although efforts are underway to identify markers linked to simple and complex traits in blueberry, marker-assisted selection is not yet a reality in blueberry breeding programs today. This is unfortunate since blueberry is especially suitable for improvement via molecular breeding because of its long generation times, high heterozygosity, inbreeding depression, and polyploidy, all of which tend to complicate genetic analyses, and can hamper traditional breeding efforts. Major savings in time, labor, and land resources could be achieved if potentially low-value genotypes could be eliminated at the seedling stage before field planting (Qu and Hancock 1997).

Efforts to optimize transformation methodologies for blueberry have been made by several groups (Rowland 1990; Graham et al. 1996; Cao et al. 1998, 2003; Song and Sink 2004, 2006; Song et al. 2007). Rowland (1990) conducted studies to investigate the susceptibility of highbush blueberry to several different *Agrobacterium tumefaciens* strains, T37, C58, A281, A518, and B6. Gall formation occurred in response to infection with three of the strains, T37, C58, and A281. Graham et al. (1996) reported transformation of blueberry half-high "Northcountry" using disarmed *A. tumefaciens* strain LBA4404 with a binary vector carrying an intron-containing GUS marker gene (Vancanneyt et al. 1990). Plants were regenerated from explants in the absence of antibiotic selection. Regenerants were shown to be GUS positive, but Southern analysis was not conducted to confirm transformation.

Cao et al. (1998) investigated several factors that can influence the early stages of transformation in blueberry. They tested disarmed *A. tumefaciens* strains LBA4404 (pAL4404) (Hoekema et al. 1983) (slightly virulent) and EHA105 (pEHA105) (Hood et al. 1993) (highly virulent), both containing the binary vector p35SGUSint, on 10 highbush cultivars. Strain EHA 105 was significantly more effective for transformation than strain LBA4404. Furthermore, four days of co-cultivation with strain EHA105 yielded 50-fold more GUS-expressing zones than two days of co-cultivation. Significant differences among cultivars were observed for both GUS-expressing leaf zones and calluses, and, for some cultivars, explant age affected the number of GUS-expressing leaf zones and calluses. Later, Cao et al. (2003) examined the effects of sucrose concentration on shoot proliferation and transfer of an intron-containing GUS gene into leaf explants from the propagated shoots. Highest GUS-expressing leaf zones were from shoots cultured on either 15 or 29 mM sucrose followed by four-days of co-cultivation with strain EHA105.

In 2004, Song and Sink reported the successful, stable transformation of four cultivars of blueberry ("Aurora", "Bluecrop", "Brigitta Blue", and "Legacy") using the *Agrobacterium* strain EHA105 containing the binary vector pBISN1, with the neomycin phosphotransferase gene (nptII) and an intron-interrupted GUS gene under the direction of the chimeric super promoter (Aocs)$_3$AmasPmas. Co-cultivation with *Agrobacterium* was for 6 days on modified woody plant medium plus 100 μM acetosyringone, and kanamycin resistant shoots were selected. Regenerants were shown to be

GUS positive and transformation was confirmed by PCR and Southern hybridizations.

Later, using their *Agrobacterium*-mediated transformation protocol (Song and Sink 2004, 2005, 2006), four chimeric bialaphos resistance (*bar*) genes driven by different promoters were evaluated by Song et al. (2007 and 2008) for production of herbicide-resistant highbush blueberry plants (*V. corymbosum* cv. Legacy). When the *bar*s were used as selectable marker genes, different promoters yielded different transformation frequencies. Three chimeric *bar* genes with the promoter nopaline synthase (nos), *Cauliflower mosaic virus* (CaMV) 35S, or CaMV 34S, yielded transgenic plants; whereas, a synthetic (Aocs)$_3$AmasPmas super promoter did not lead to successful regeneration of transgenic plants. Three month-old plants of three separate transgenic events each for the 35S and *nos* promoters, as well as non-transgenic plants, were sprayed with the herbicide glufosinate ammonium (GS) at five levels (mg·L^{-1}: 0, 750, 1,500, 3,000 and 6,000). Evaluations of leaf damage two weeks after spraying indicated that all transgenic plants exhibited much higher herbicide resistance than non-transgenic plants. Additionally, the transgenic plants with the *35S-bar* showed a higher herbicide resistance than those with the *nos-bar*. After application of eight times the standard level of GS (6,000 mg·L^{-1}) applied in the field, over 90% of the leaves with the *35S-bar* and 19.5%–51.5% of the leaves with the *nos-bar* showed no symptom of herbicide damage; whereas, only 5% of the non-transgenic leaves had no damage. When one-year-old, field-grown plants were sprayed with 750 mg·L^{-1} GS, transgenic plants of four transgenic events with the *nos-bar* survived with variations in the level of foliar damage; in contrast, all non-transgenic plants died.

1.11 Structural Genomics

Approximately 5,000 blueberry ESTs are currently publicly available in the EST database of GenBank. The ESTs were generated from four different flower bud cDNA libraries, two standard and two subtracted libraries. These libraries were constructed as part of a genomics-based research project to identify genes associated with cold acclimation in blueberry.

The two standard cDNA libraries were constructed using RNA from cold acclimated (CA) and non-acclimated (NA) floral buds of the cv. Bluecrop (Levi et al. 1999; Dhanaraj et al. 2004). "Bluecrop" was chosen because it is the industry standard and is fairly cold hardy. As part of this project, ~ 1,200 5'-end ESTs were generated from each library and ~ 100 3' end ESTs were generated from the CA library (Dhanaraj et al. 2004, 2007). The 2,480 high quality ESTs from the two libraries were assembled into contigs or clusters based on the presence of overlapping, identical, or similar sequences (Dhanaraj et al. 2007). From the contig analysis, 1,527 clusters were formed. These included 458 singletons from the CA library and 615 from the NA library, 204 contigs comprised of only CA ESTs, 116 contigs comprised of only NA ESTs, and 134 contigs comprised of ESTs from both libraries. Thus, only 8.8% (134/1,527) of the total distinct transcripts were shared between the two libraries, suggesting marked differences in gene expression under the two conditions. Individual ESTs and assembled contigs were compared with the National Center for Biotechnology Information (NCBI) non-redundant protein database using the BLASTX algorithm (Altschul et al. 1997). In this way, about 57% of the ESTs from both libraries were assigned putative functions based on sequence similarity to genes or proteins of known function in the Genbank. Of the remaining sequences, 27% showed significant similarity to protein or DNA sequences that were of unknown function and 16% had no significant similarity to any other sequences in the databases.

Later, forward and reverse subtracted libraries were also prepared from "Bluecrop" flower buds. This was in order to identify important regulatory genes that are often expressed at rather low levels and over a short time frame that might have been missed by random picking and sequencing of clones from the standard cDNA libraries. A forward subtracted library (SL) was prepared in such a way to enrich for transcripts that are expressed at higher levels at 400 hours of cold acclimation than at 0 hours of cold acclimation and vice versa for a reverse subtracted library (RL) (Naik et al. 2007). Approximately 500 ESTs from the SL and ~ 170 ESTs from the RL were generated. Contig analyses and BLAST searches were performed to categorize the genes. From a contig analysis of all the ESTs from both libraries, 254 singletons and 118 contigs were formed, representing a total of 372 distinct transcripts. Of the 118 contigs, 50 included sequences from the SL only, 57 included sequences from the RL only, and 11 contigs included sequences from both libraries. Again, a very low percentage, ~ 3% or 11/372, of the total distinct transcripts were shared between the libraries.

The most highly abundant cDNAs that were picked from each of the four libraries are listed in Table 1-2. Those that represent a much higher percentage of clones picked from one library than from the opposite library (shown in parentheses in the table) represent potentially differentially expressed transcripts. Many of the genes from the standard libraries were confirmed to be cold-responsive by Northern blot analyses (Dhanaraj et al. 2004). For example, of the cDNAs that were more abundant in the CA library than in the NA library, those encoding a dehydrin, probable cytochrome P450 monooxygenase, early light-inducible protein, and β-amylase were all confirmed to be induced with low temperature exposure. Of the cDNAs that were more abundant in the NA library than in the CA library, those encoding a histone H3.2 protein and BURP-domain dehydration-responsive protein RD 22 were confirmed to be suppressed with low temperature exposure

Table 1-2 Most abundant cDNAs from the various blueberry EST libraries (CA = standard cold acclimated, NA = standard non-acclimated, SL = forward subtracted, RL = reverse subtracted). The putative gene identities and the frequency of each of the cDNAs in their respective libraries are given. The frequency of each of the cDNAs in the opposite library (CA vs. NA, or SL vs. RL) is provided in parentheses for comparison.

Putative gene identification	Frequency of cDNAs
Cold acclimated library (CA)	
Probable cytochrome P450 monooxygenase	1.8% (0% in NA)
Dehydrin	1.1% (0.8% in NA)
F1 ATPase subunit α	0.8% (0.5% in NA)
Early light-inducible protein	0.5% (0.1% in NA)
DNA J heat shock protein	0.5% (0% in NA)
B-amylase	0.5% (0.2% in NA)
Non-acclimated library (NA)	
BURP-domain dehydration-responsive protein RD 22	1.1% (0.1% in CA)
Dehydrin	0.8% (1.1% in CA)
Metallothionein-like protein	0.7% (0.1% in CA)
Histone H3.2/H3.3 protein	0.5% (0.1% in CA)
F1 ATPase subunit α	0.5% (0.8% in CA)
Major latex-like protein	0.5% (0% in CA)
Forward subtracted library (SL)	
Early light-inducible protein	30.8% (0% in RL)
β-amylase	2.8% (0% in RL)
Dehydrins	2.6% (0% in RL)
Ribulose 1,5 bisphosphate carboxylase/oxygenase small subunit	2.2% (1.8% in RL)
Late Embryogenesis Abundant (LEA) proteins	1.8% (0% in RL)
Galactinol synthase	1.6% (0% in RL)
Proline-rich proteins	1.6% (0% in RL)
NADH dehydrogenase chain	1.2% (0% in RL)
Zinc finger proteins	0.8% (0.6% in RL)
Extensin	1.0% (0% in RL)
Nodulin-24	1.0% (0% in RL)
Seed maturation family proteins	0.8% (0% in RL)
Ion transporters	0.8% (0% in RL)
bZIP family transcription factors	0.6% (0% in RL)
F1 ATPase subunit alpha	0.6% (0% in RL)
Reverse subtracted library (RL)	
AP2-domain proteins	3.6% (0% in SL)
Anthocyanidin reductases	3.0% (0% in SL)
Protein kinases	2.4% (0.2% in SL)
Mitochondrial uncoupling proteins	1.8% (0% in SL)
GDSL-motif lipase/hydrolase	1.8% (0.2% in SL)
Ribulose 1,5 bisphosphate carboxylase/oxygenase small subunit	1.8% (2.2% in SL)
Chitinase	1.8% (0.2% in SL)

(Dhanaraj et al. 2004). Several of the genes from the subtracted libraries were confirmed to be differentially expressed, as well, using quantitative RT-PCR (Naik et al. 2007). These included genes encoding several zinc finger proteins from the SL as well as genes encoding certain AP2-domain proteins and anthocyanidin reductase from the RL. Furthermore, many potential regulatory genes were identified from the subtracted libraries such as calmodulin, putative myb-related protein, putative bZIP protein, zinc finger proteins, AP2 domain-containing proteins like CBF, and a basic-helix-loop-helix transcription factor, among others.

An aliquot of the NA standard floral bud library (Dhanaraj et al. 2004) was also provided to scientists working on the multi-institutional "Floral Genomic Project" (*http://pgn.cornell.edu*). Researchers generated another 1,758 5' end ESTs from this library. From a contig analysis of these sequences, 1,549 unigenes were assembled into 138 contigs and 1,411 singletons.

1.12 Functional Genomics: Transcriptomics

All the functional genomic studies in blueberry to date have focused on the biology of cold tolerance with an ultimate goal of applying the information learned to develop more cold hardy cultivars. These studies, unlike those with *Arabidopsis*, have focused on cold acclimation of flower bud tissue, rather than leaf tissue, because damage to flower buds directly results in reduction in fruit yield. Initially, using a molecular genetic approach, several cold-responsive genes were identified from blueberry flower buds including 65, 60, and 14 kD dehydrins (Muthalif and Rowland 1994). Dehydrins are a group of heat-stable, glycine-rich plant proteins that are induced by environmental stimuli that have a dehydrative component including drought, low temperature, salinity, and seed maturation (Close 1996). Good quantitative correlation between dehydrin accumulation in blueberry flower buds and cold hardiness was found in several different genotypes studied (Muthalif and Rowland 1994; Arora et al. 1997). Full-length cDNA clones encoding the 60 kD (Levi et al. 1999) and the 14 kD (Dhanaraj et al. 2005) dehydrins were isolated and the seasonal expression of their messages were found to be similar to their protein accumulation patterns.

Because of the complexity and multigenic nature of cold tolerance, many other genes are likely involved in cold acclimation in blueberry in addition to dehydrins. Therefore, more recently, researchers have constructed cDNA microarrays to identify genes that are differentially expressed during cold acclimation. About 2,500 clone inserts from the standard CA and NA libraries, from which ESTs (described above) were generated, were used to construct the cDNA microarrays. The microarrays were used to examine changes in abundance of gene transcripts in floral buds of the blueberry cv. Bluecrop at multiple times during cold acclimation under field (0, 70, 400, 800, and 1,200 chill units) and cold room (0, 500, and 1,000 chill units) conditions (Dhanaraj et al. 2007). Both field and cold room conditions were studied because there are advantages (and disadvantages) to both systems. First, the purpose of this research was to identify genes potentially involved not just in cold stress but in cold acclimation. In woody perennials, cold acclimation is triggered by several environmental cues, including short photoperiod, low temperature, available moisture, etc. Second, field plants have the advantage over cold-treated greenhouse plants of acclimating to cold under natural conditions, to progressively shorter photoperiods and colder temperatures, and to reaching greater cold hardiness levels. Floral buds of "Bluecrop", in particular, reach a maximum cold hardiness level (temperature causing 50% injury or LT_{50}) that is 2–3°C lower under field conditions than in the cold room (Arora et al. 1997; Muthalif and Rowland 1994; Rowland et al. 2005). Third, one is not as limited by the number of plants that can be cold acclimated, or therefore the amount of flower buds that can be collected, in the field as compared with the cold room. On the other hand, the cold room has the advantage of being a more controlled or "controllable" environment than the field. Other stresses may be acting in the field that would not be a factor in the cold room. Because of these considerations, both systems were used in this study and the results were compared.

Lists of all the genes induced or suppressed at any of the time points from these microarray studies are provided in supplemental tables in Dhanaraj et al. (2007). Many of the cold induced genes identified in *Arabidopsis* (Seki et al. 2001; Fowler and Thomashow 2002; Hannah et al. 2005) were identified as being highly induced in CA blueberry under both field and cold room conditions, such as galactinol synthase, beta amylase, LEAs, dehydrins, and early light-inducible protein. In addition, other genes were identified as being cold-induced in blueberry under field and cold room conditions that have not been reported as cold-induced in *Arabidopsis*, such as auxin-repressed protein, protein kinase PINOID, pectate lyase-like protein, and S-adenosylmethionine decarboxylase proenzyme. Of particular interest is PINOID, which is involved in auxin-mediated signaling. In *Arabidopsis*, PINOID has been shown to work in concert with LEAFY to direct formation of the floral meristem (Ezhova et al. 2000; Lebedeva et al. 2005). In blueberry, it appears that PINOID is induced by low temperature. These differences between blueberry and *Arabidopsis* may reflect differences between woody perennials and herbaceous annuals or differences in the types of tissues analyzed (flower buds versus leaves).

Although many of the same genes were induced and suppressed under field and cold room conditions in blueberry, there were also many differences in the types of genes expressed under the two conditions. In general, the number of cold-induced genes was lower and the number of cold-suppressed genes was higher in the field than in the cold room. The total number of differentially induced genes during cold acclimation in the field was 134 or 5.2% of cDNAs that were arrayed. The number of genes induced under cold room conditions was 241 or 9.4% of the arrayed genes, which was equivalent to almost twice the number induced under field conditions. Many of the genes induced under cold room conditions that were not induced under field conditions could be divided into three major groups: 1) genes associated with stress tolerance; 2) those that encode glycolytic and TCA cycle enzymes; and 3) those that encode protein synthesis machinery. Examples of some of the encoded proteins associated with stress tolerance are phospholipid hydroperoxide glutathione peroxidase, glutathione-S-transferase, metallothionein-like protein type 2, and low temperature-induced 78 kD protein. The glycolytic and TCA cycle enzymes, whose messages were upregulated under cold room conditions but not field conditions, include phosphoglyceromutase, glyceraldehyde-3-phosphate dehydrogenase, malate dehydrogenase, and succinate dehydrogenase flavoprotein alpha subunit. The protein synthesis machinery members include many 30S, 40S, and 60S ribosomal proteins, and translation initiation factor 5A-4. The upregulation of genes involved in respiration and protein synthesis may simply relate to the observation that more genes are upregulated under cold room conditions than under field conditions, thus the need for more energy and more translation factors.

The total number of differentially suppressed genes during cold acclimation in the field was 162 or 6.3% of arrayed genes. In contrast, the number of genes suppressed in the cold room was much lower, only 32 or 1.2% of arrayed genes. It is interesting that many of the genes that were suppressed during cold acclimation in the field were, in fact, induced during cold acclimation in the cold room. Examples include genes encoding ubiquitin-conjugating enzyme, heat shock proteins, phospholipid hydroperoxide glutathione peroxidase, and low temperature-induced 78 kD protein. Most of the genes suppressed in the field reached maximum suppression at the 400 hour time point. An examination of the weather data revealed that, in the week preceding this collection point, the average temperature had dropped below freezing, to –2.1°C from an average temperature of 6.3 °C the week before. Thus, it was hypothesized that many of these genes may be suppressed by freezing temperatures rather than by low temperature, possibly explaining why they were not suppressed in the cold room experiment.

From this work it was concluded that, although there are similarities in the types of genes that respond during cold acclimation in the cold room and in the field environment, there are major differences. There may be a general response to turn on many stress-related genes at 4 °C, but once temperatures drop below freezing, many of these genes are turned off, perhaps to save energy. This also indicates that many of these stress-related genes are not, in fact, required for reaching maximum cold tolerance levels, since "Bluecrop" field plants, which have been acclimated to winter conditions, are slightly more cold hardy than cold room plants.

Currently, transcription profiles from "Bluecrop" are being compared to those of a more cold-sensitive genotype, "Tifblue" (Rowland et al. 2008a). Although the analysis is not yet complete, preliminary results indicate that some transcripts that are highly induced by cold in "Bluecrop" are not induced in "Tifblue". Also, some transcripts that are induced by cold in both "Bluecrop" and "Tifblue", are induced earlier and to higher levels in "Bluecrop". For example, transcripts of galactinol synthase and "blind", an abiotic stress protein identified in potato, were both induced more than two-fold in "Bluecrop" but were not induced in "Tifblue". Transcripts of a dehydrin and a putative cell wall protein were turned on earlier, stayed on longer, and were induced to higher levels in "Bluecrop" than in "Tifblue" buds. In this way, by comparing transcript profiles of cold hardy and cold-sensitive genotypes, the list of cold-responsive genes can be narrowed down to what might be the key players in the cold acclimation pathway.

1.13 Role of Bioinformatics as a Tool

Recently, the first publicly available, online database for blueberry genomics data was created (Alkharouf et al. 2007). The Blueberry Genomics Database (BBGD); http://bioinformatics.towson.edu/BBGD/was established to serve blueberry and, as it expands, the broader *Ericaceae* communities. Presently, its primary focus is to store and analyze ESTs and microarray data generated from experiments aimed at studying cold acclimation and mid-winter hardiness of blueberry (described above). It allows for the correlation of gene function, deduced from the EST data, with expression levels, deduced from the microarray data.

The microarray and EST data are available in easy-to-search formats. Regarding the microarray data, users can choose to query a specific time point in the cold acclimation experiment, across all or selected time points, or query by gene name, ID, or GenBank accession number. Users can also conduct advanced queries to find genes that have similar expression in different experiments and/or biological samples. Results from cluster analysis and online analytical processing are also displayed. Regarding the EST data, results of EST analysis and contig assembly for the EST libraries, along with graphical representations and charts, are available. Like the microarray portion, users can query the sequence database by gene name, ID, or GenBank accession number. Users can also browse a specific library in a table format.

An example of how the EST database is currently being used is in the ongoing development of EST-PCR markers (Rowland et al. 2003c) for mapping, DNA-fingerprinting, and genetic relationship studies of blueberry. Because a goal of some of the mapping projects is to identify QTL for cold hardiness, efforts to map as many of the cold-responsive genes identified from the microarray study as is possible, since they are good candidate genes for controlling cold hardiness, are underway.

1.14 Conclusions and Future Prospects

There is no doubt that blueberry genomics is still in its infancy. Currently, all the publicly available ESTs are from cDNA libraries made from flower buds (cold acclimated and non-acclimated). This needs to be expanded to include leaf, fruit, root, and stem libraries at the minimum and other abiotic (mineral soils, heat, drought) and biotic (disease) stress libraries as well. The maps of both diploid and tetraploid populations need to be further saturated with SSRs and EST-PCR markers, more populations segregating for other traits need to be developed, and markers identified that are associated with the various traits of interest. More gene expression studies need to be undertaken including more microarrays to sort out genes that are expressed in response to various stimuli and during development. Interesting genes need to be identified and tested in transformations with *Arabidopsis*, or other model systems, and then blueberry. Efforts to improve the transformation efficiency of blueberry should continue. Work in blueberry proteomics and metabolomics should begin perhaps initially to improve fruit nutritional quality. Available gene sequences should be utilized to better define the phylogenetic relationships among the different *Vaccinium* species. Efforts to coordinate a whole genome sequencing project should be made.

References

Aalders LE, Hall IV (1962) The inheritance of white fruit in the velvet-leaf blueberry, *Vaccinium myrtilloides* Michx. Can J Genet Cytol 4: 90–91.

Aalders LE, Hall IV (1963) The inheritance and taxonomic significance of the "nigrum" factor in the common lowbush blueberry, *Vaccinium angustifolium*. Can J Genet Cytol 5: 115–118.

Alkharouf NW, Dhanaraj AL, Naik D, Overall C, Matthews BF, Rowland LJ (2007) BBGD: an online database for blueberry genomic data. Biomed Central Plant Biol 7: 5.

Altschul SF, Madden TL, Schaffer AA, Zhang J, Zhang A, Miller W, Lipmann DJ (1997) Gapped BLAST and PSI-BLAST: a new generation of protein database search programs. Nucl Acids Res 25: 3389–3402.

Arce-Johnson P, Rios M, Zuniga M (2002) Identification of blueberry varieties using random amplified polymorphic DNA markers. Acta Hort 574: 221–224.

Arora R, Rowland LJ, Panta GR (1997) Chill responsive dehydrins in blueberry: are they associated with cold hardiness or dormancy transitions? Physiol Plant 101: 8–16.

Arora R, Rowland LJ, Panta GR, Lim C-C, Lehman JS, Vorsa N (1998) Genetic control of cold hardiness in blueberry. In: PH Li , THH Chen (eds) Plant Cold Hardiness: Molecular Biology, Biochemistry, and Physiology. Plenum Publishing Corp, New York, USA, pp 99–106.

Arora R, Rowland LJ, Lehman JS, Lim CC, Panta GR, Vorsa N (2000) Genetic analysis of freezing tolerance in blueberry (*Vaccinium* section *Cyanococcus*). Theor Appl Genet 100: 690–696.

Arora R, Rowland LJ, Ogden EL, Dhanaraj AL, Marian CO, Ehlenfeldt MK, Vinyard B (2004) Dehardening kinetics, bud development, and dehydrin metabolism in blueberry (*Vaccinium* spp.) cultivars during deacclimation at constant, warm temperatures. J Am Soc Hort Sci 129: 667–674.

Aruna M, Ozias-Akins P, Austin ME, Kochert G (1993) Genetic relatedness among rabbiteye blueberry (*Vaccinium ashei*) cultivars determined by DNA amplification using single primers of arbitrary sequence. Genome 36: 971–977.

Aruna M, Austin ME, Ozias-Akins P (1995) Randomly amplified polymorphic DNA fingerprinting for identifying rabbiteye blueberry (*Vaccinium ashei* Reade) cultivars. J Am Soc Hort Sci 120: 710–713.

Ballinger WE, Maness EP, Galletta GJ, Kushman LJ (1972) Anthocyanins of ripe fruit of a "pink-fruited" hybrid of highbush blueberries, *Vaccinium corymbosum* L. J Am Soc Hort Sci 97: 381–384.

Ballington JR (1990) Germplasm resources available to meet future needs for blueberry cultivar improvement. Fruit Var J 44: 54–62.

Ballington JR (2001) Collection, utilization and preservation of genetic resources in *Vaccinium*. HortScience 36: 213–220.

Ballington JR, Galletta GJ (1976) Potential fertility levels in four diploid *Vaccinium* species. J Am Soc Hort Sci 101: 507–509.

Ballington JR, Kirkman WB, Ballinger WE, Maness EP (1988) Anthocyanin, aglycone and aglycone-sugar content in the fruits of temperate North American species of four sections in *Vaccinium*. J Am Soc Hort Sci 113: 746–749.

Ballington JR, Rooks SD, Milholland RD, Cline WO, Meyer JR (1993) Breeding blueberries for pest resistance in North Carolina. Acta Hort 346: 87–94.

Bassil NV, Oda A, Hummer K (2009) Blueberry microsatellite markers identify cranberry cultivars. Acta Hort 810: 181–186.

Bell DJ, Rowland LJ, Polashock JJ, Drummond FA (2008) Suitability of EST-PCR markers developed in highbush blueberry for genetic fingerprinting and relationship studies in lowbush blueberry and related species. J Am Soc Hort Sci 133: 701–707.

Bittenbender HC, Howell GS (1975) Interactions of temperature and moisture content on spring deacclimation of flower buds of highbush blueberry. Can J Plant Sci 55: 447–452.

Boches PS, Bassil NV, Rowland LJ (2005) Microsatellite markers for *Vaccinium* from EST and genomic libraries. Mol Ecol Notes 5: 657–660.

Boches P, Bassil NV, Rowland LJ (2006a) Genetic diversity in the highbush blueberry *Vaccinium corymbosum* L. evaluated with microsatellite markers. J Am Soc Hort Sci 131: 674–686

Boches P, Rowland LJ, Hummer K, Bassil NV (2006b). Cross-species amplification of SSR loci in the genus *Vaccinium*. Acta Hort 715: 119–128.

Brevis P, Hancock J, Rowland LJ (2007) Development of a genetic linkage map for tetraploid highbush blueberry using SSR and EST-PCR markers. HortScience 42: 963.

Brevis PA, Bassil NV, Ballington JR, Hancock JF (2008) Impact of wide hybridization on highbush blueberry breeding. J Am Soc Hort Sci 133: 1–11.

Brooks SJ, Lyrene PM (1998a) Derivatives of *Vaccinium arboreum* x *Vaccinium* Section *Cyanococcus*: I. Morphological characteristics. J Am Soc Hort Sci 123: 273–277.

Brooks SJ, Lyrene PM (1998b) Derivatives of *Vaccinium arboreum* x *Vaccinium* Section *Cyanococcus*: II. Fertility and fertility parameters. J Am Soc Hort Sci 123: 997–1003.

Bruederle LP, Vorsa N (1994) Genetic differentiation of diploid blueberry, *Vaccinium* sect. *Cyanococcus* (*Ericaceae*). Syst Bot 19: 337–349.

Bruederle LP, Vorsa N, Ballington JR (1991) Population genetic structure in diploid blueberry *Vaccinium* section *Cyanococcus* (*Ericaceae*). Am J Bot 78: 230–237.

Burgher K, Jamieson A, Lu X, McRae K (1998) Genetic relationships of 26 wild blueberry clones (*Vaccinium angustifolium*), as determined by random amplified polymorphic DNA analysis. In: Proc of the Atlantic Plant Tissue Culture Assoc Annu Meet. Fredericton, New Brunswick, Canada, p 7.

Burgher KL, Jamieson AR, Lu X (2002) Genetic relationships among lowbush blueberry genotypes as determined by randomly amplified polymorphic DNA analysis. J Am Soc Hort Sci 127: 98–103.

Camp WH (1945) The North American blueberries with notes on other groups of *Vacciniaceae*. Brittonia 5: 203–275.

Cao X, Liu Q, Rowland LJ, Hammerschlag FA (1998) GUS expression in blueberry (*Vaccinium* spp.): factors influencing *Agrobacterium*-mediated gene transfer efficiency. Plant Cell Rep 18: 266–270.

Cao X, Fordham I, Douglass L, Hammerschlag F (2003) Sucrose level influences micropropagation and gene delivery into leaves from *in vitro* propagated highbush blueberry shoots. Plant Cell Tiss Org Cult 75: 255–259.

Caruso FL, Ramsdell DC (1995) Compendium of Blueberry and Cranberry Diseases. APS Press, St. Paul, MN, USA.

Chandler CK, Draper AD, Galletta GJ (1985) Crossability of a diverse group of polyploidy interspecific hybrids. J Am Soc Hort Sci 110: 878–881.

Clark JR, Moore JN, Draper AD (1986) Inheritance of resistance to *Phytophthora* root rot in highbush blueberry. J Am Soc Hort Sci 111: 106–109.

Cline WO, Schilder A (2006) Identification and control of blueberry diseases. In: NF Childers , PM Lyrene (eds) Blueberries for Growers, Gardeners, Promoters. Childers Publ, Gainesville, Florida, USA, pp 115–138.

Close TJ (1996) Dehydrins: emergence of a biochemical role of a family of plant dessication proteins. Physiol Plant 97: 795–803.

Cockerman LE, Galletta GJ (1976) A survey of pollen characteristics in certain *Vaccinium* species. J Am Soc Hort Sci 101: 671–676.

Connor AM, Luby JJ, Finn CE, Hancock JF (2002a) Genotypic and environmental variation in antioxidant activity, total phenolics and anthocyanin content among blueberry cultivars. J Am Soc Hort Sci 127: 89–97.

Connor AM, Luby JJ, Tong JJ (2002b) Variability in antioxidant activity in blueberry and correlations among different antioxidant assays. J Am Soc Hort Sci 127: 238–244.

Czesnik E (1985) Investigation of F_1 generation of interspecific hybrids *Vaccinium corymbosum* L. x *V. uliginosum* L. Acta Hort 165: 85–91.

Darrow GM (1960) Blueberry breeding, past, present, future. Am Hort Mag 39: 14–33.

Darrow GM, Scott DH, Derman H (1954) Tetraploid blueberries from hexaploid x diploid species crosses. Proc Am Soc Hort Sci 63: 266–270.

Dhanaraj AL, Slovin JP, Rowland LJ (2004) Analysis of gene expression associated with cold acclimation in blueberry floral buds using expressed sequence tags. Plant Sci 166: 863–872.

Dhanaraj AL, Slovin JP, Rowland LJ (2005) Isolation of a cDNA clone and characterization of expression of the highly abundant, cold acclimation-associated 14 kDa dehydrin of blueberry. Plant Sci 168: 949–957.

Dhanaraj AL, Alkharouf NW, Beard HS, Chouikha IB, Matthews BF, Wei H, Arora R,Rowland LJ (2007) Major differences observed in transcript profiles of blueberry during cold acclimation under field and cold room conditions. Planta 225: 735–751.

Draper AD (1977) Tetraploid hybrids from crosses of diploid, tetraploid and hexaploid *Vaccinium* species. Acta Hort 61: 33–36.

Draper AD, Scott DH (1971) Inheritance of albino seedlings in tetraploid highbush blueberry. J Am Soc Hort Sci 96: 791–792.

Draper AD, Hancock J (2003) Florida 4B: native blueberry with exceptional breeding value. J Am Pom Soc 57: 138–141.

Draper AD, Stretch AM, Scott DH (1972) Two tetraploid sources of resistance for breeding blueberries resistant to *Phytophthora cinnamomi* Rands. HortScience 7: 266–268.

Ehlenfeldt MK (2001) Self and cross-fertility in recently released highbush cultivars. HortScience 36: 133–135.

Ehlenfeldt MK, Stretch AW (2000) Mummy berry blight resistance in rabbiteye blueberry cultivars. HortScience 35: 1326–1328.

Ehlenfeldt MK, Prior RL (2001) Oxygen radical absorbance capacity (ORAC) and phenolic and anthocyanin concentrations in fruit and leaf tissues of highbush blueberry. J Agri Food Chem 49: 2222–2227.

Ehlenfeldt MK, Rowland LJ (2006) Cold-hardiness of *Vaccinium ashei* and *V. constablaei* germplasm and the potential for northern-adapted rabbiteye cultivars. Acta Hort 715: 77–80.

Ehlenfeldt MK, Vorsa N (2007) Inheritance patterns of parthenocarpic fruit development in highbush blueberry (*Vaccinium corymbosum* L.). HortScience 42: 1127–1130.

Ehlenfeldt MK, Stretch AW, Draper AD (1993) Sources of genetic resistance to red ringspot virus in a breeding population. J Am Soc Hort Sci 28: 207–208.

Ehlenfeldt MK, Draper AD, Clark JR (1995) Performance of southern highbush blueberry cultivars released by the U.S. Department of Agriculture and cooperating state agriculture stations. HortTechnol 5: 127–130.

Ehlenfeldt MK, Rowland LJ, Arora R (2003) Bud hardiness and deacclimation in blueberry cultivars with varying species ancestry: flowering time may not be a good indicator of deacclimation. Acta Hort 626: 39–44.

Ehlenfeldt MK, Ogden EL, Rowland LJ, Vinyard B (2006) Evaluation of mid-winter cold hardiness among 25 rabbiteye blueberry (*Vaccinium ashei* Reade) cultivars. HortScience 41: 579–581.

Ehlenfeldt MK, Rowland LJ, Ogden EL, Vinyard B (2007) Bud cold hardiness of *Vaccinium ashei, V. constablaei*, and hybrid derivatives and their potential for producing northern-adapted rabbiteye cultivars. HortScience 42: 1131–1134.

Erb W, Draper AD, Swartz HJ (1990) Combining ability for plant and fruit traits of interspecific blueberry progenies on mineral soil. J Am Soc Hort Sci 115: 1025–1028.

Erb W, Draper AD, Swartz HJ (1993) Relation between moisture stress and mineral soil adaptation in blueberries. J Am Soc Hort Sci 118: 130–134.

Erb W, Draper AD, Swartz HJ (1994) Combining ability for seedling root system size and shoot vigor in interspecific blueberry progenies. J Am Soc Hort Sci 119: 793–797.

Ezhova TA, Soldatova OP, Kalinina AI, Medvedev SS (2000) Interaction of ABRUPTUS/PINOID and LEAFY genes during floral morphogenesis in *Arabidopsis thaliana* (L.) Heynh. Genetika 36: 1682–1687.

Finn CE, Luby JJ (1986) Inheritance of fruit development interval and fruit size in blueberry progenies. J Am Soc Hort Sci 11: 784–788.

Finn CE, Luby JJ, Rosen CJ, Ascher PD (1993a) Blueberry germplasm screening at several soil pH regimes. I. Plant survival and growth. J Am Soc Hort Sci 118: 377–382.

Finn CE, Rosen CJ, Luby JJ, Ascher PD (1993b) Blueberry germplasm screening at several soil pH regimes. II. Plant nutrient composition. J Am Soc Hort Sci 118: 383–387.

Finn CE, Hancock JF, Mackey T, Serce S (2003) Genotype x environment interactions in highbush blueberry (*Vaccinium* sp. L.) families grown in Michigan and Oregon. J Am Soc Hort Sci 128: 196–200.

Fowler S, Thomashow MF (2002) *Arabidopsis* transcriptome profiling indicates that multiple regulatory pathways are activated during cold acclimation in addition to the CBF cold response pathway. Plant Cell 14: 1675–1690.

Galletta GJ (1975) Blueberries and cranberries. In: J Janick , JN Moore (eds) Advances in Fruit Breeding. Purdue Univ Press, West Lafayette, Indiana, USA, pp 154–196.

Galletta GJ, Ballington JR (1996) Blueberries, cranberries and lingonberries. In: J Janick , JN Moore (eds) Fruit Breeding, vol II. Vine and Small Fruit Crops. John Wiley and Sons, Inc., New York, USA, pp 1–107.

Graham J, Greig K, McNicol RJ (1996) Transformation of blueberry without antibiotic selection. Ann Appl Biol 128: 557–564.

Haghighi K, Hancock JF (1992) DNA restriction fragment length variability in genomes of highbush blueberry. HortScience 27: 44–47.

Hall IV, Aalders LE (1963) Two-factor inheritance of white fruit in the common blueberry, *Vaccinium angustifolium* Ait. Can J Genet Cytol 5: 371–373.

Hancock JF, Schulte NL, Siefker JH, Pritts MP, Roueche JM (1982) Screening highbush blueberry cultivars for resistance to the aphid *Illinoia pepperi*. HortScience 17: 362–363.

Hancock JF, Morimoto KM, Schulte NL, Martin JM, Ramsdell DC (1986) Search for resistance to blueberry shoestring virus in highbush blueberry cultivars. Fruit Var J 40: 56–58.

Hancock JF, Nelson JW, Bittenbender HC, Callow PW, Cameron JS, Krebs SL, Pritts MP, Schumann CM (1987) Variation among highbush blueberry cultivars in susceptibility to spring frost. J Am Soc Hort Sci 112: 702–706.

Hancock JF, Draper AD (1989) Blueberry culture in North America. HortScience 24: 551–556.

Hancock JF, Sakin S, Callow PW (1991) Heritability of flowering and harvest dates in *Vaccinium corymbosum*. Fruit Var J 45: 173–176.

Hancock JF, Erb WA, Goulart BL, Scheerens JC (1997) Blueberry hybrids with complex backgrounds evaluated on mineral soils : cold hardiness as influenced by parental species and location. Acta Hort 446: 389–396.

Hancock JF, Luby JJ, Beaudry R (2003) Fruits of the Ericaceae. In: L Trugo , P Fingas , B Caballero (eds) Encyclopedia of Food Science, Food Technology and Nutrition. Academic Press, London, UK, pp 565–599.

Hancock JF, Lyrene P, Finn CE, Vorsa N, Lobos GA (2008) Blueberry and cranberries. In: JF Hancock (ed) Temperate Fruit Crop Breeding: Germplasm to Genomics. Springer Science+Business Media B.V., Berlin, Germany, pp 115–149.

Hannah MA, Heyer AG, Hincha DK (2005) A global survey of gene regulation during cold acclimation in *Arabidopsis thaliana*. PLoS Genet 1: 179–196.

Hanson EJ, Berkheimer SF, Hancock JF (2007) Seasonal changes in the cold hardiness of the flower buds of highbush blueberry with varying species ancestry. J Am Pom Soc 61: 14–18.

Hiirsalmi H (1977) Inheritance of characters in hybrids of *Vaccinium uliginosum* and highbush blueberries. Ann Agri Fenn 16: 7–18.

Hill NM, Vander Kloet SP (1983) Zymotypes in *Vaccinium* section *Cyanococcus* and related groups. Proc Nova Scotian Inst of Sci 33: 115–121.

Hoekema A, Hirsch PR, Hooykaas PJJ, Schilperoort RA (1983) A binary plant vector strategy based on separation of vir- and T-regions of the *Agrobacterium tumefaciens* Ti-plasmid. Nature 303: 179–180.

Hokanson K, Hancock J (1998) Levels of allozymic diversity in diploid and tetraploid *Vaccinium* sect. *Cyanococcus* (blueberries). Can J Plant Sci 78: 327–332.

Hood EE, Gelvin SB, Melchers LS, Hoekema A (1993) New *Agrobacterium* vectors for plant transformation. Transgen Res 2: 208–218.

Jarne P, Lagoda PJL (1996) Microsatellites, from molecules to population and back. Trends Ecol Evol 11: 424–429.

Jelenkovic G (1973) Breeding value of pentaploid interspecific hybrids of *Vaccinium*. Jugoslovensko Vocarstvo 7: 237–244.

Johnstons (1946) Observations on hybridizing lowbush and highbush blueberries. Proc Am Soc Hortic Sci 47: 199–200.

Krebs SL, Hancock JF (1989) Tetrasomic inheritance of isoenzyme markers in the highbush blueberry *Vaccinium corymbosum* L. Heredity 63: 11–18.

Kron KA, Powell EA, Luteyn JL (2002) Phylogenetic relationships within the blueberry tribe (*Vaccinieae, Ericaceae*) based on sequence data from *mat*K and nuclear ribosomal ITS regions, with comments on placement of *Satyria*. Amer J Bot 89: 327–336.

Lebedeva OV, Ondar UN, Penin AA, Ezhova TA (2005) Effect of the ABRUPTUS/PINOID gene on expression of the LEAFY gene in *Arabidopsis thaliana*. Genetika 41: 559–565.

Levi A, Rowland LJ (1997) Identifying blueberry cultivars and evaluating their genetic relationships using randomly amplified polymorphic DNA (RAPD) and simple sequence repeat- (SSR-) anchored primers. J Am Soc Hort Sci 122: 74–78.

Levi A, Rowland LJ, Hartung JS (1993) Production of reliable randomly amplified polymorphic DNA (RAPD) markers from DNA of woody plants. HortScience 28: 1188–1190.

Levi A, Panta GR, Parmentier CM, Muthalif MM, Arora R, Shanker S, Rowland LJ (1999) Complementary DNA cloning, sequencing, and expression of an unusual dehydrin from blueberry floral buds. Physiol Plant 107: 98–109.

Luby JJ, Finn CE (1986) Quantitative inheritance of plant growth habit in blueberry progenies. J Am Soc Hort Sci 111: 609–611.

Luby JJ, Finn CE (1987) Inheritance of ripening uniformity and relationship to crop load in blueberry progenies. J Am Soc Hort Sci 112: 167–170.

Luby JJ, Ballington JR, Draper AD, Pliszka K, Austin ME (1991) Blueberries and cranberries (*Vaccinium*). In: JN Moore , JR Ballington (eds) Genetic resources of temperate fruit and nut crops. Int Soc for Hort Sci, Wageningen, The Netherlands, pp 391–456.

Lyrene PM (1985) Effects of year and genotype on flowering and ripening dates in rabbiteye blueberry. HortScience 20: 407–409.

Lyrene PM (1988) An allele for anthocyanin-deficient foliage, buds, and fruit in *Vaccinium elliottii*. J Hered 79: 80–82.

Lyrene PM (1991) Fertile derivatives from sparkleberry x blueberry crosses. J Am Soc Hort Sci 116: 899–902.

Lyrene PM (2008) Breeding southern highbush blueberries. In: J Janick (ed) Plant Breeding Reviews, vol 30. John Wiley and Sons, New York, USA, pp 354–406.

Lyrene PM, Ballington JR (1986) Wide hybridization in *Vaccinium*. HortScience 21: 52–57.

Lyrene PM, Sherman WB (1983) Mitotic instability and 2n gamete production in *Vaccinium corymbosum* x *V. elliottii* hybrids. J Am Soc Hort Sci 108: 339–342.

Lyrene PM, Vorsa N, Ballington JR (2003) Polyploidy and sexual polyploidization in the genus *Vaccinium*. Euphytica 133: 27–36.

Martin RR, Bristow PR, Wegener LA (2006) Scorch and shock: emerging virus diseases of highbush blueberry and other *Vaccinium* species. Acta Hort 715: 463–467.

Meyer JR, Ballington JR (1990) Resistance of *Vaccinium* species to the leafhopper *Scaphytopius magdalensis* (Homoptera: Cicadellidae). Ann Entomol Soc Am 83: 515–520.

Moerman DE (1998) Native American Ethnobotany. Timber Press, Portland, Oregon, USA.

Moore JN (1993) The blueberry industry of North America. Acta Hort 346: 15–26.

Munoz CE, Lyrene PM (1985) In vitro attempts to overcome the cross-incompatibility between *V. corymbosum* L. and *V. elliottii* Chapm. Theor Appl Genet 69: 591–596.

Muthalif MM, Rowland LJ (1994) Identification of dehydrin-like proteins responsive to chilling in floral buds of blueberry (*Vaccinium*, section *Cyanococcus*). Plant Physiol 104: 1439–1447.

Naik D, Dhanaraj AL, Arora R, Rowland LJ (2007) Identification of genes associated with cold acclimation in blueberry (*Vaccinium corymbosum* L.) using a subtractive hybridization approach. Plant Sci 173: 213–222.

Ortiz R, Vorsa N, Bruederle LP, Laverty T (1992) Occurrence of unreduced pollen in diploid blueberry species, *Vaccinium* sect. *Cyanococcus*. Theor Appl Genet 85: 55–60.

Panta GR, Rieger MW, Rowland LJ (2001) Effect of cold and drought stress on blueberry dehydrin accumulation. J Hort Sci Biotechnol 76: 549–556.

Perry JL, Lyrene PM (1984) In vitro induction of tetraploidy in *Vaccinium darrowi, V. elliottii*, and *V. darrowi* x *V. elliottii* with colchicine treatment. J Am Soc Hort Sci 109: 4–6.

Polashock JJ, Kramer M (2006) Resistance of blueberry cultivars to *Botryosphaeria* stem blight and *Phomopsis* twig blight. HortScience 41: 1457–1461.

Polashock J, Vorsa N (2006) Segregating blueberry populations for mummy berry fruit rot resistance. New Jersey Annu Veg Meeting Proc, Atlantic City, New Jersey, USA.

Polashock JJ, Ehlenfeldt MK, Stretch AW, Kramer M (2005) Anthracnose fruit rot resistance in blueberry cultivars. Plant Dis 89: 33–38.

Powell LE (1987) Hormonal aspects of bud and seed dormancy in temperate-zone woody plants. HortScience 22: 845–850.

Quamme HA (1985) Winter hardiness research in North America on woody perennials. Acta Hort 168: 191–194.

Qu L, Hancock JF (1995) Nature of 2n gamete formation and mode of inheritance in interspecific hybrids of diploid *Vaccinium darrowi* and tetraploid *V. corymbosum*. Theor Appl Genet 91: 1309–1315.

Qu L, Hancock JF (1997) Randomly amplified polymorphic DNA- (RAPD-) based genetic linkage map of blueberry derived from an interspecific cross between diploid *Vaccinium darrowi* and tetraploid *V. corymbosum*. J Am Soc Hort Sci 122: 69–73.

Qu L, Vorsa N (1999) Desynapsis and spindle abnormalities leading to 2n pollen formation in *Vaccinium darrowi*. Genome 42: 35–40.

Qu L, Hancock JF, Whallon JH (1998) Evolution in an autopolyploid group displaying predominantly bivalent pairing at meiosis: Genomic similarity of diploid *Vaccinium darrowi* and autotetraploid *V. corymbosum* (Ericaceae). Am J Bot 85: 698–703.

Ranger CM, Johnson-Cicalese J, Polavarapu S, Vorsa N (2006) Evaluation of *Vaccinium* spp. for *Illinoia pepperi* (Hemiptera: Aphididae) performance and phenolic content. J Econ Entomol 99: 1474–1482.

Ranger CM, Singh AP, Johnson-Cicalese J, Polavarapu S, Vorsa N (2007) Intraspecific variation in aphid resistance and constitutive phenolics exhibited by the wild blueberry *Vaccinium darrowi*. J Chem Ecol 33: 711–729.

Rousi A (1963) Hybridization between *Vaccinium uliginosum* and cultivated blueberry. Ann Agri Fenn 2: 12–18.

Rowland LJ (1990) Susceptibility of blueberry to infection by *Agrobacterium tumefaciens*. HortScience 25: 1659.

Rowland LJ, Levi A (1994) RAPD-based genetic linkage map of blueberry derived from a cross between diploid species (*Vaccinium darrowi* and *V. elliottii*). Theor Appl Genet 87: 863–868.

Rowland LJ, Hammerschlag FA (2005) *Vaccinium* spp. Blueberry. In: RE Litz (ed) Biotechnology of Fruit and Nut Crops. CABI Publ, Cambridge, Massachusetts, USA, pp 222–246.

Rowland LJ, Lehman JS, Levi A, Ogden EL, Panta GR (1998) Genetic control of chilling requirement in blueberry. In: WO Cline, JR Ballington (eds) Proc 8th North Am Blueberry Res and Extn Worker's Conf, North Carolina State Univ, Raleigh, North Carolina, USA, pp 258–267.

Rowland LJ, Ogden EL, Arora R, Lim C-C, Lehman JS, Levi A, Panta GR (1999) Use of blueberry to study genetic control of chilling requirement and cold hardiness in woody perennials. HortScience 34: 1185–1191.

Rowland LJ, Dhanaraj AL, Polashock JJ, Arora R (2003a) Utility of blueberry-derived EST-PCR primers in related *Ericaceae* species. HortScience 38: 1428–1432.

Rowland LJ, Mehra S, Dhanaraj A, Ogden EL, Arora R (2003b) Identification of molecular markers associated with cold tolerance in blueberry. Acta Hort 625: 59–69.

Rowland LJ, Mehra S, Dhanaraj AL, Ogden EL, Slovin JP, Ehlenfeldt MK (2003c) Development of EST-PCR markers for DNA fingerprinting and genetic relationship studies in blueberry (*Vaccinium*, section *Cyanococcus*). J Am Soc Hort Sci 128: 682–690.

Rowland LJ, Panta GR, Mehra S, Parmentier-Line C (2004) Molecular genetic and physiological analysis of the cold-responsive dehydrins of blueberry. J Crop Improv 10: 53–76.

Rowland LJ, Ogden EL, Ehlenfeldt MK, Vinyard B (2005) Cold hardiness, deacclimation kinetics, and bud development among 12 diverse blueberry genotypes under field conditions. J Am Soc Hort Sci 130: 508–514.

Rowland LJ, Dhanaraj AL, Naik D, Alkharouf N, Matthews B, Arora R (2008a) Study of cold tolerance in blueberry using EST libraries, cDNA microarrays, and subtractive hybridization. HortScience 43: 1975–1981.

Rowland LJ, Ogden EL, Ehlenfeldt MK, Arora R (2008b) Cold tolerance of blueberry genotypes throughout the dormant period from acclimation to deacclimation. HortScience 43: 1970–1974.

Sakai A, Larcher W (1987) Frost Survival of Plants: Responses and Adaptation to Freezing Stress. Springer, Berlin, Heidelberg, New York.

Scheerens JC, Erb WA, Goulart BL, Hancock JF (1993a) Blueberry hybrids with complex genetic backgrounds evaluated on mineral soils: stature, growth rate, yield potential and adaptability to mineral soil conditions as influenced by parental species. Fruit Var J 53: 73–90.

Scheerens JC, Erb WA, Goulart BL, Hancock JF (1993b) Blueberry hybrids with complex genetic backgrounds evaluated on mineral soils: flowering, fruit development, yield and yield components as influenced by parental species. Fruit Var J 53: 91–104.

Seki M, Narusaka M, Abe H, Kasuga M, Yamaguchi-Shinozaki K, Carninci P, Hayashizaki Y, Shinozaki K (2001) Monitoring the expression pattern of 1300 Arabidopsis genes under drought and cold stresses by using a full-length cDNA microarray. Plant Cell 13: 61–72.

Sharp RH, Darrow GM (1959) Breeding blueberries for the Florida climate. Proc Fla State Hort Soc 72: 308–311.

DE, Soltis PS (1993) Molecular data and the dynamic nature of polyploidy. Crit Rev Plant Sci 12: 243–273.

Soltis DE, Soltis PS (1999) Polyploidy: recurrent formation and genome evolution. Trends Ecol Evol 14: 348–352.

Soltis DE, Soltis P, Schemske DW, Hancock JF, Thompson JN, Husband BC, Judd WS (2007) Autopolyploidy in angiosperms: have we grossly underestimated the number of species. Taxon 56: 13–30.

Song G-Q, Sink KC (2004) *Agrobacterium tumefaciens*-mediated transformation of blueberry (*Vaccinium corymbosum* L.). Plant Cell Rep 23: 475–484.

Song G-Q, Sink KC (2005) Plant regeneration and transformation of blueberry. Recent Res Dev Genet Breed Signpost 2: 190–200.

Song G-Q, Sink KC (2006) Blueberry (*Vaccinium corymbosum* L.). In: K Wang (ed) Agrobacterium Protocols.-Methods in Molecular Biology Book Series. 2nd edn. Humana Press, Totowa, New Jersey, USA, pp 263–272.

Song G-Q, Roggers RA, Sink KC, Particka M, Zandstra B (2007) Production of herbicide-resistant highbush blueberry 'Legacy' by *Agrobacterium*-mediated transformation of the *bar* gene. Acta Hort 738: 397–407.

Song G-Q, Sink KC, Callow PW, Baughan R, Hancock JF (2008) Evaluation of different promoters for production of herbicide-resistant blueberry plants. J Am Soc Hort Sci 133: 605–611.

Stommel JR, Panta GR, Levi A, Rowland LJ (1997) Effects of gelatin and BSA on the amplification reaction for generating RAPD. BioTechniques 22: 1064–1066.

Stretch AW, Ehlenfeldt MK (2000) Resistance to the fruit infection phase of mummy berry disease in highbush blueberry cultivars. HortScience 35: 1271–1273.

Stretch AW, Ehlenfeldt MK, Brewster V (1995) Mummy berry blight resistance in highbush blueberry cultivars. HortScience 30: 589–591.

Strik B (2005) Blueberry—an expanding world berry crop. Chron Hort 45: 7–12.

Strik BC, Yarborough D (2005) Blueberry production trends in North America, 1992 to 2003, and predictions for growth. HortTechnol 15: 391–398.

Vancanneyt G, Schmidt R, O'Connor-Sanchez A, Willmitzer L, Rocha-Sosa M (1990) Construction of an intron-containing marker gene: splicing of the intron in transgenic plants and its use in monitoring early events in *Agrobacterium*-mediated plant transformation. Mol Gen Genet 220: 245–250.

Vander Kloet SP (1980) The taxonomy of highbush blueberry, *Vaccinium corymbosum*. Can J Bot 58: 1187–1201.

Vander Kloet SP (1988) The genus *Vaccinium* in North America. Res Branch Agri Can Publ 1828. 201 pp.

Varshney RK, Graner A, Sorrells ME (2005) Genic microsatellite markers in plants: features and applications. Trends Biol 23: 48–55.

Vorsa N (1997) On a wing: the genetics and taxonomy of *Vaccinium* from a pollination perspective. Acta Hort 446: 59–66.

Vorsa N, Ballington JR (1991) Fertility of triploid highbush blueberry. J Am Soc Hort Sci 116: 336–341.

Vorsa N, Rowland LJ (1997) Estimation of $2n$ megagametophyte heterozygosity in a diploid blueberry (*Vaccinium darrowi* Camp) clone using RAPDs. J Hered 88: 423–426.

Vorsa N, Jelenkovic G, Draper AD, Welker WV (1987) Fertility of 4x x 5x and 5x x 4x progenies derived from *Vaccinium ashei/corymbosum* pentaploid hybrids. J Am Soc Hort Sci 112: 993–997.

Wang Z, Weber JL, Zhong G, Tanksley SD (1994) Survey of plant short tandem DNA repeats. Theor Appl Genet 88: 1–6.

Weber JL, May PE (1989) Abundant class of human DNA polymorphisms which can be typed using the polymerase chain reaction. Am J Hum Genet 44: 388–396.

Weiser CJ (1970) Cold resistance and injury in woody plants. Science 169: 1269–1278.

Welsh J, McClelland M (1990) Fingerprinting genomes using PCR with arbitrary primers. Nucl Acids Res 18: 7213–7218.

Wiedow C, Buck EJ, Gardiner S, Scalzo J (2007) Genetic diversity within a blueberry (*Vaccinium* spp. and hybrids) germplasm collection at HortResearch. In: Plant Anim Genome XV Conf, San Diego, CA, USA, p 468.

Williams JGK, Kubelik AR, Livak KJ, Rafalski JA, Tingey SV (1990) DNA polymorphisms amplified by arbitrary primers are useful as genetic markers. Nucl Acids Res 18: 6531–6536.

Wisniewski M, Bassett C, Gusta LV (2003) An overview of cold hardiness in woody plants: seeing the forest through the trees. HortScience 38: 952–959.

2

Cranberry

Anna Zdepski,[1,2] Samir C. Debnath,[3] Amy Howell,[2,4] James Polashock,[5] Peter Oudemans,[2,4] Nicholi Vorsa[2,4] and Todd P. Michael[1,2,]*

2.1 Introduction

2.1.1 History of the Cranberry

The American cranberry, *Vaccinium macrocarpon* Aiton ($2n = 2x = 24$), is native to North America. A member of the Ericaceae (Heath) Family, within the subsection *Oxycoccus*, cranberry is adapted to moist, acidic soils, peat bogs, marshes, and swamps with a temperate climate. The natural distribution ranges from Newfoundland west through the Great Lakes region to western Ontario and Minnesota, and south at higher elevations in the Appalachian Mountains to North Carolina and Tennessee (Vander Kloet 1988; Fig. 2.1). As a commercial crop it has been introduced to a wider range and can be found between longitudes 70 W and 80 W and latitudes 40 N and 50 N as well as in introductions to Britain, western Canada, and western United States. It is also cultivated in eastern Europe in Latvia, Estonia, and Belarus and in South America in Chile.

V. macrocarpon is a low-growing woody perennial that reproduces sexually via flowers and asexually through stolons. The creeping evergreen stolons support ascending shoots that produce flowers and subsequent

[1]The Waksman Institute of Microbiology, Rutgers, The State University of New Jersey, Piscataway, NJ, USA.
[2]Department of Plant Biology and Pathology, The School of Environmental and Biological Sciences, Rutgers, The State University of New Jersey, New Brunswick, NJ, USA.
[3]Atlantic Cool Climate Crop Research Centre, Agriculture and Agri-Food Canada, P.O. Box 39088, 308 Brookfield Road, St. John's, Newfoundland and Labrador A1E 5Y7, Canada.
[4]Philip E. Marucci Center for Blueberry and Cranberry Research, Rutgers, The State University of New Jersey, Chatsworth, NJ, USA.
[5]Genetic Improvement of Fruits and Vegetables Lab, USDA-ARS, Chatsworth, NJ, USA.

*Corresponding author: *tmichael@waksman.rutgers.edu*

berries. Flowers have an inferior ovary and after pollination, primarily by bees, the ovary and calyx fuse to form a true berry. Cavities (locules) containing seeds (Fig. 2-2) allow ripe fruit to float in water, providing an effective method of "water harvesting" (Fig. 2-3). The berry epidermis develops a deep red color from concentrated anthocyanins as they ripen. Harvest, frost control, and pest prevention are facilitated on commercial farms by controlled flooding. Newly planted cranberry beds bear fruit after 2–4 years and will continue producing a crop for as many as 75 years. For most of the 20th century established beds were rarely renovated and replaced with new varieties, and many original varieties have remained in place for decades.

Figure 2-1 Native cranberry habitat in North America.

The earliest documented commercial cranberry farm was established around 1816 in Cape Cod Massachusetts, where cranberries were transplanted from native stands into prepared beds and exported to New York City for profit (Trehane 2004). Wild *V. macrocarpon* was selected by early farmers for superior horticultural characteristics such as fruit size, fruit color, and resistance to pests and disease. Over 100 varieties have been selected and named from the wild (Dana 1983). Many farmers still use wild selections, such as "Early Black", which was first cultivated in 1845. Formal cranberry breeding programs were initiated in 1929 through the support of the USDA and "Stevens", released in 1950 is the most popular "improved" variety planted today.

Table 2-1 Popular cranberry varieties.

Cultivar	Origin	Year of Introduction
"Early Black"	Selection from wild population	1835
"Ben Lear"	Selection from wild population	1901
"Stevens"	"McFarlin" x "Potter"	1950
"Franklin"	"Early Black" x "Howes"	1961

2.1.2 Other Vaccinium Species

Roughly 400 shrubs belong to the genus *Vaccinium*. They can be found throughout North America, South America, Europe, and Asia in acidic, sandy, peaty, or organic soils (Vander Kloet 1988). Twenty-five species within the genus are endemic to North America, and many have edible fruits.

The small and sparsely fruited *Vaccinium oxycoccos* var. *microcarpum* (small cranberry) grows in low arctic tundra and has a history of usage by Native People in Washington State, British Columbia, and Alaska (vander Kloet 1988) and is found in Greenland, northern Europe and northern Asia into Japan (Trehane 2004). *Vaccinium oxycoccos* does not have the same commercial value as the cultivated American cranberry, but it is picked for consumption in areas where *V. macrocarpon* cannot survive. The hardy *V. oxycoccos* grows low to the ground, and is usually associated with thick mosses that protect it from harsh winds (Trehane 2004).

Blueberries (*Vaccinium* sect. *Cyanococcus*) have several cultivated species. Two species of "wild" lowbush blueberries (*V. angustifolium* and *V. myrtilloides*) are low-growing rhizomatous shrubs with sweet berries

Figure 2-2 Ripe cranberry fruits showing seeds and locules.

Color image of this figure appears in the color plate section at the end of the book.

Figure 2-3 The cranberry harvest in New Jersey.

commercially harvested from New Hampshire north to Prince Edward Island, Nova Scotia, New Brunswick, and Newfoundland. The blueberry industry is dominated by the northern highbush blueberry *V. corymbosum* commercially cultivated for its large berry size.

Lingonberry (*Vaccinium vitis-idaea*) is also known in North America as the partridgeberry or northern mountain cranberry and grows in the cold climates of subarctic Alaska, Canada, Washington, and Oregon. It can also be found on the East Coast as far south as Connecticut. Lingonberry is cultivated or collected in the wild in northern Europe. The Canadian province, Newfoundland and Labrador, is the largest North American lingonberry producing region (Penney et al. 1997) with about 140,000 kg harvested annually from native stands for processing, mostly for export (Jamieson 2001). Smaller commercial quantities are harvested in Nova Scotia and Labrador. The main market is for juice and sauce production and consumed mostly in northern Europe. Cultivars of Lingonberry are produced from Germany, Sweden, Holland, Finland, and a breeding program at The University of Wisconsin (Trehane 2004).

2.1.3 Biology of Cranberry

In nature, *V. macrocarpon* reproduces both sexually through seed and asexually through stolons. Stolon sections in contact with soil root readily. Ascending shoots, colloquially referred to as "uprights", are produced along the length of the stolon and are often terminated by a floral inflorescence bud, which overwinters. Typically, floral inflorescence buds are initiated in late summer and early fall. After receiving chilling (> 1,000 hours), the inflorescence bud breaks dormancy in mid to late spring forming an inflorescence acropetally of 3–7 flowers borne singly in axils of reduced leaves along the rachis and terminating in a leafy shoot flowering in early summer. Flowers are 4-merous, perfect, having eight anthers with an inferior four-locule ovary (> 20 ovules) and style about 1 cm in length. Flowers are protandrous, with the style 6–7 mm in length at anthesis inside the anther whorl, then elongating to 8–10 mm extending 2–3 mm beyond the anther whorl 2–3 days post anthesis (Fig. 2-4). The stigma produces an exudate and appears most receptive 3-5 days after anthesis, producing an exudate. Characteristic of the family Ericaceae, cranberry pollen is shed as

Figure 2-4 Cranberry flowers.

Color image of this figure appears in the color plate section at the end of the book.

a tetrad with the four pollen grains of a meiotic event held in a tetrahedral formation. All four pollen grains of the tetrad are potentially viable. Anthers of a single cranberry flower may shed 7–28 thousand pollen tetrads (Cane et al. 1996). Typically, one to a few fruit(s) is/are set upright, but varietal variation may play a role.

Commercial beds are established by pressing or "disking-in" dormant vines into the bed media in early spring. Vine material can either be stolons from pruning or mowing existing commercial beds. These propagation methods have resulted in problems with varietal genetic homogeneity in numerous commercial beds since "off-types" typically produce more vegetative tissue (i.e., stolons) and are thus inadvertently propagated. A more recent approach has been the utilization of rooted cuttings in a nursery system with DNA-fingerprinted and virus indexed material. Rooted cuttings are planted at a density of one per square foot or greater.

2.1.4 Disease of Cranberry

The cultivated cranberry is vulnerable to attack by a broad diversity of fungal plant pathogens. The most economically important is a group of these pathogensascomycete fungi that collectively cause fungal fruit rot. Over 15 fungal species are known to infect the cranberry fruit and incite fruit rot crop loss (Fig. 2-5). New Jersey growing conditions promote the greatest fruit rot pressure of any cranberry growing areas. Some of the more common fungal fruit rot species include: *Colletotrichum acutatum, C. gloeosporioides, Pestalotia Physalospora vaccinii* (blotch rot), and *Phyllosticta vaccinii* (early rot), and *Coleophoma empetri* (ripe rot). Other species that can occasionally be important include *Fusicoccum putrefaciens* (end rot), *Phomopsis vaccinii* (viscid rot), *Physalospora vaccinii* (blotch rot), *Allantophomopsis cytisporea* (black rot), *A. lycopodina* (black rot), *Strasseria geniculata* (black rot), and *Fusicoccum putrefaciens* (end rot) and *Coleophoma empetri* (ripe rot). Most fruit rotting organisms also infect vegetative tissues, providing a source of overwintering inoculum. Post-harvest fungal rots are significant in fresh market cranberries, occurring during storage, with the most important diseases being end rot and black rot. Recently, a macroarray method has been developed to rapidly and accurately identify these cranberry fruit rotting organisms without isolation and culturing on artificial medium (Robideau et al. 2008).

Without adequate fungicide controls, fruit rot in New Jersey and Massachusetts can reach 100% within one or two years (Oudemans et al. 1999). Even with modern chemical control, losses due to fruit rot in the

Figure 2-5 A diseased cranberry plant exhibiting necrotic tissues.
Color image of this figure appears in the color plate section at the end of the book.

northeast growing regions (New Jersey and Massachusetts) range from 1 to 60% in a given year. In Wisconsin, incidence of fruit rot is more sporadic and varies widely from year to year and bed to bed (McManus et al. 2003). In the western growing regions (Oregon, Washington, and British Columbia) losses due to fruit rot tend to be lower than those in the northeast (Oudemans et al. 1999). Surveys have suggested that although the contribution of any given fungal species can vary from year to year and location to location, there appears to be no strong correlation between fungal species prevalence and variety (Stiles and Oudemans 1999). Rather, differences in susceptibility are probably a combination of 1) biochemical, such as the production of antifungal compounds, 2) physical, such as fruit surface waxes, or 3) phenological, such as avoidance of pathogens due to variation in bloom time and fruit ripening. All of these characteristics are heritable and as such, may be used to enhance resistance through breeding.

Cotton ball, caused by the fungus *Monilinia oxycocci*, is a different type of fruit rotting pathogen. It is the most important cranberry disease in Wisconsin and is also found in Massachusetts and western growing regions. This highly specialized fungus causes shoot blight as a result of ascospore infections. Conidia produced on infected shoots infect fruit by growing down the stylar canal and filling the fruit locules with a cottonly white mycelium. Infected fruits are unmarketable and in some years, losses can be significant. No differences in susceptibility have been detected among the commonly grown cultivars (McManus et al. 1999).

Fairy ring is a disease of increasing importance. First described in the 1930s, it was considered to be of minor importance. As cultivation practices have evolved and the practice of sanding cranberry beds has gained acceptance, this disease has become a real threat and is the most difficult and costly to control. The disease develops as expanding circular patches with the advancing edge characterized by rapidly dying plants. As patches merge, the pattern on affected beds becomes a series of arcs of dying vines. The disease causes a direct crop loss as well as providing an entry point for invasive weeds into the cranberry bed and also causes an increase in genetic diversity of the cranberry crop presumably by influencing the deposition and germination of seed. These effects can greatly reduce the productive life of a cranberry bed. To date resistance to this disease is unknown, however it is most common in the cultivar "Ben Lear" (Oudemans et al. 2008).

The phytoplasma disease "false-blossom" devastated the cranberry industry in New Jersey in the early 1900s. As such, the first cranberry breeding program was initiated to develop false-blossom disease resistant cultivars (Chandler et al. 1947, 1950). This disease causes flowers to deform and thus produce little or no fruit. The phytoplasma is transmitted by the blunt-nosed leafhopper and therefore resistance breeding was focused on selection for "non-feeding" preference of this insect vector. The cultivar

"Shaw's Success" was considered to be the most resistant (least preferable by the leaf-hopper), while "Early Black" was moderately resistant and "Howes" were considered to be susceptible (more preferable by the blunt-nosed leaf hopper). Current use of effective insecticides has limited the widespread damage associated with this disease during the 1900s.

Another serious disease of cultivated cranberry is *Phytophthora* root and runner rot. It is caused by the oomycete *Phytophthora cinnamomi* and at least one other species (Polashock et al. 2005). Found in the surface waters used to irrigate cranberries, *Phytophthora* species are widespread pathogens (Oudemans 1999). *Phytophthora* can impact crop yields by both acute and chronic phases of the disease. In its chronic phase the plant growth and yield are reduced but, plants continue to grow and symptoms may go unnoticed unless inspected closely. In the acute phase infected plants die suddenly due to extensive root damage, usually during hot, dry periods (Pozdnyakova et al. 2002). Resistance to *Phytophthora cinnamomi* has been observed in cranberry germplasm, it is currently unknown if tolerance to this root rot is heritable.

There are also a variety of stem and leaf pathogens of cranberry that are generally less important than either fruit rot or *Phytophthora* root rot. Losses, when they occur, may be severe but are usually sporadic. The most common among these is upright dieback caused is by *Phomopsis vaccinii*. Another stem disease, cranberry stem gall, is caused by several species of auxin-producing soil borne bacteria, which are likely introduced via wounding during the harvesting process (Vasanthakumar and McManus 2004). Powdery mildew caused by *Microsphaera vaccinii* is common during greenhouse propagation and occasionally in the field especially when excess nitrogen is applied.

Viruses are rare in cranberry. To date, only three viruses have been reported in cranberry and these include *Ringspot virus, Blueberry scorch virus,* and *Tobacco streak virus* (Boone 1995; Teifon Jones et al. 2001; Wegener et al. 2004). These are largely uncharacterized in cranberry and seem to have little effect on plant health or productivity. Only ringspot is known to show any effect and that is limited to red spots on the leaves of some varieties and superficial red rings on the fruit.

2.1.5 Current Market and Industry

Cranberries are cultivated commercially in the United States and Canada. Wisconsin has the largest cranberry industry, producing roughly one half of all US cranberries followed in production by Massachusetts, New Jersey, Oregon and Washington.

Cranberry production in the US was reported to be 6.40 million barrels (313 million kilos) in 2007. Most berries are destined for processing in the juice industry with seasonal fresh fruit making up only 5% of the market in 2003 (USDA 2008). Canned cranberry sauce was first produced in Maine in 1888 and has remained a Thanksgiving seasonal favorite.

Cranberry farmers have historically organized into cooperatives to stabilize the industry through uniform prices. Today the two major cooperatives in the US are Ocean Spray Cranberries and Northland Cranberries. Ocean Spray began as an agricultural cooperative of growers in 1912 and became a national company in 1930. In 2005 the cooperative estimated sales at US $1.4 billion dollars.

Cranberry is one of three major endemic North American fruit crops to be cultivated for commercial sale, along with the blueberry and Concord grape. Four cultivars of cranberry form the basis for commercial production: "Early Black" selected in 1845, "Howes" selected in 1843, "McFarlin" selected in 1874, and "Searles" selected in 1893 "Stevens" from a USDA cross selected in 1950 is a common choice for new plantings (Trehane 2004). The economic pressure to replace beds of "Early Black" and "Howes" with newly released, higher yield cultivars has lead to planting of "Grygleski#1", "Crimson Queen", "Mullica Queen", and "Demoranville".

2.1.6 Current Research—Natural Products and Chemical Analysis—Cranberry Health Benefits

The recent excitement over the biological activity of cranberry components has elevated the cranberry's widespread appeal to consumers. The cranberry accumulates some of the highest concentrations of phenolic compounds among fruit species commonly consumed in the United States (Vvedenskaya et al. 2004). The panoply of nutritional qualities and health benefits has led to the cranberry being called a "super-food".

Consumption of cranberry juice has been associated with prevention of urinary tract infections, and several well-designed clinical studies have confirmed this association (Avorn et al. 1994; Kontiokari et al. 2001). For years, the preventative effects of cranberry were assumed to be due to the acidity of the fruit; however a majority of clinical studies have demonstrated that a bacteriostatic pH is rarely achieved in urine following normal serving sizes of cranberry (Avorn et al. 1994; Walker et al. 1997). Urinary tract infection is initiated by bacterial adherence to the uroepithelium, followed by bacterial multiplication and colonization of the urinary tract (Beachy 1981). Cranberries contain a group of polyphenolic compounds called proanthocyanidins with double A-type linkages that inhibit P-fimbriated *E. coli* from adhering to the uroepithelial cells (Foo et al. 2000a, b; Howell et al. 1998), thus preventing growth and subsequent infection. These proanthocyanidins are different from those found in other foods, such as chocolate and grape that contain single B-type linkages between flavan-3-ol units (Gu et al. 2003).

Cranberry proanthocyanidins are associated with preventing bacterial adhesion not only in the urinary tract, but also in the stomach and oral cavity. They prevented *Helicobacter pylori*, the bacterium that causes stomach ulcers, from attaching to isolated stomach cells (Burger et al. 2000). In a randomized placebo-controlled clinical study in China, consumption of two 250-ml servings of cranberry juice cocktail (27% cranberry juice) per day accounted for a 15% eradication of *H. pylori* (Zhang et al. 2005). Cranberry proanthocyanidin extracts have also demonstrated activity in the oral cavity preventing biofilms associated with coaggregation and adhesion of bacteria to teeth and gums (Weiss et al. 2002), and prevented adhesion of *Streptoccocus sobrinus* to hydroxyapatite (Steinberg et al. 2004). Cranberry polyphenolics inhibit glucosyl transferases in glucan synthesis, and ATPase activities required by *Streptococcus mutans*, the main pathogen in dental caries, reducing acidogenicity in biofilm establishment (Duarte et al. 2006; Gregoire et al. 2007).

Cranberry consumption is also showing promise for reducing risk factors for heart disease and inhibition of cancer cell growth. Cranberries are very high in both antioxidants and anti-inflammatory compounds (Neto 2007), both of which may mediate heart disease and cancer. Cranberry has been shown to inhibit LDL cholesterol oxidation, induce LDL receptor expression and increase cholesterol uptake (Chu and Liu 2005; McKay and Blumberg 2007). Clinically, daily consumption of 250 ml of cranberry juice cocktail increased high density lipoprotein (HDL) cholesterol, which is associated with heart disease risk reduction, by over 8% in overweight men (Ruel and Couillard 2007). Studies on cranberry's chemopreventive properties include inhibition of breast tumor growth(Guthrie 2000), inhibition of acid-induced cell proliferation in human esophageal adenocarcinoma cells (Kresty et al. 2008), and inhibition of MCF-7 and MDA-MB-435 breast cancer cell growth (Ferguson et al. 2004). Since reactive oxygen species have negative effects on a number of physiological processes in the body and are known to increase cancer risk, the high antioxidant capacity of cranberry proanthocyanidins, flavonol glycosides and anthocyanins may be responsible for the anti-carcinogenic properties (Neto 2007).

Breeding cranberries for increased antioxidant levels has been investigated by Vorsa and Johnson-Cicalese with the conclusions that genetic variation and environmental influences contribute to natural variation of flavonols, anthocyanins, and proanthocyanidins. Anthocyanins are produced in the fruit epidermis and are the pigments that give cranberry its red color. The highest concentration of anthocyanins is produced during fruit ripening. Flavonols and proanthocyanidins are produced in highest concentrations during flowering and early fruit development. These quantities decline as the fruit matures, but levels increase slightly just as the cranberries become ripe (Vvedenskaya et al. 2004). It is postulated that the increase of flavonols and proanthocyanidins may provide protection against fungal pathogens and herbivory during flowering and fruit set (Harborne 1997). The high levels of flavonols in cranberry plants may serve as protection from UV light, oxidative damage, and as general pathogen and predator deterrent (Vvedenskaya et al. 2004).

2.2 Cranberry Genetic Resources

2.2.1 *Natural Collections*

Cranberry genetic resources are maintained in field plots in major cranberry growing areas either in experimental stations or in commercial beds. The commercial selections are exposed to risks from contamination from other genotypes, damage by severe weather, pests, and disease.

For successful cranberry crop development, breeders need a wide range of gene sources. The maintenance of genetic resources is the focus of the United States National Plant Germplasm System, and a collection of 200 genotypes of cranberries are preserved as seed or maintained as clones in the genebank at the National Clonal Germplasm Repository in Oregon. A collection of cultivars and 455 wild cranberry clones is also located in St. John's, Canada as part of the Atlantic Cool Climate Crop Research Centre. The wild genotypes were selected from native Canadian bogs for their horticultural characteristics and are maintained for research, breeding, and germplasm preservation.

The cranberry has a high amount of inherent and unexploited genetic diversity. Working with 13 wild cranberry clones collected from Newfoundland and Labrador, Canada, researchers reported a significant variation among the clones with the similarity values ranging from 0.37 to 0.91 (Debnath 2005). In another study, researchers evaluated 43 wild cranberry clones collected from four Canadian provinces (Nova Scotia, New Brunswick, Prince Edward Island, and Quebec) and five cranberry cultivars using Randomly Amplified Polymorphic DNA-based strategies (Debnath 2007a). Cluster analysis by the unweighted pair-group method with arithmetic averages and the principal coordinate analysis grouped the wild clones and cultivars into a number of clusters and sub-clusters. Geographical distribution explained 10% of total variation as revealed by analysis of molecular variance. Reasons for the apparent lack of a geographical differentiation could be the result of a glacial bottleneck and rapid colonization coupled with autogamous breeding habit of cranberries (Stewart and Nilsen 1995). This low level of differentiation could also reflect high rates of regional gene flow resulting from both human migration and agricultural trade (Aldrich and Doebley 1992). These postulations are supported by population studies employing isozymes, which also indicated low genetic diversity (Bruederle et al. 1996).

2.2.2 Genetic Fingerprinting of Cranberry Cultivars

The procumbent growth habit of cranberry leads to dense mat beds with individual plants difficult to delineate in mature plantings. Seeds from fallen fruit may germinate and establish themselves within a mature bed, resulting in cultivar contamination. Varietal misidentification is furthered by the lack of a regulated nursery industry for supply of cranberry cultivars. Planting material is sold with little verification of varietal identity and purity.

The chance of loss of cultivar clonal integrity is great, especially if the cultivars are propagated for many years. Contamination sources include a mixed planting in the original bed, the establishment of volunteer seedlings among the clonal cultivar, and the accumulation of mutants over time. Redundancies of identical or near-identical accessions may occur in germplasm collections due to documentation errors, collection from genetically homogenous sites, and exchange of identical accessions between gene banks. Identification and elimination of redundancies is an important aspect of plant genetic resource management. Accurate and rapid genotype identification is especially important in vegetatively propagated plant species for germplasm characterization, practical breeding purposes, and proprietary-right protection.

Understanding the distribution of genetic diversity among individuals, populations and gene pools is crucial for the efficient management of germplasm collections and breeding programs. Diversity analysis is routinely carried out using sequencing of selected gene(s) or molecular marker technologies.

2.2.3 Crosses and Interesting Genetics

V. macrocarpon is diploid with the basic genome $n = 12$ and is represented by 12 linkage groups. The size of the diploid *V. macrocarpon* genome is estimated to be 470 megabases (Table 2-2). Disomic inheritance is evident with a few qualitative markers such as a leaf anthocyanin deficiency trait, and is consistent with Mendelian inheritance. Cranberry appears to be moderately to highly self-fertile. The major cultivars are largely self-fertile, however cross-pollination yields slightly more developed seeds/berry, and percent developed seeds than self-pollination (Sarracino and Vorsa 1991).

Backcrosses, e.g., "Franklin" ("Early Black" x "Howes") x Early Black and Wilcox ("Howes" x "Searles") x "Howes" having inbreeding coefficients of F = 0.25 did not exhibit significant reduction in developed seed number as compared to outcrosses. In fact, up to five selfing generation

Table 2-2 Comparative genome sizes of cranberry and other plant species.

Species	Common Name	Diploid Genome Size (Mb)
Arabidopsis thaliana	Mustard Weed	157
Brachypodium distachyon	Purple False Brome	300
Carica papaya	Papaya	372
Oryza sativa	Rice	430
Vaccinium darrowii	Evergreen Blueberry	444
Vaccinium macrocarpon	**Cranberry**	**470**
Vitis vinifera	Grape	500
Populus trichocarpa	Poplar	550
Sorghum bicolor	Sorghum	680
Glycine max	Soybean	1,115
Zea mays	Maize	2,500

(S_5) lineages have been produced from a number of cultivars including "Pilgrim", "Stevens" and "Ben Lear", suggesting cranberry tolerates F > 0.97 inbreeding levels, with no gross inviability effects. Reciprocal translocation heterozygotes such as found in the cultivar Howes, with a recessive lethal system, may maintain linkage blocks interstitial and proximal to the breakpoint in an advantageous heterozygous state, in an inbreeding population (Ortiz and Vorsa 2004). Interestingly, a proportion of selfed progeny from "Howes" and "Wilcox" are completely sterile. The asexual reproduction phase of cranberry would maintain these genotypes in natural populations, and may actually be advantageous with most resources directed toward stolon production.

Qualitative traits are few in cranberry. A leaf anthocyanin deficient variety, referred to as "Murphy's Green", was identified, and inheritance data from first generation hybrids, backcrosses, and first generation self-pollinations indicate a monogenic recessive trait. A fruit anthocyanin deficient variety, "Yellow Bell", was also identified in a native Maine population. Open-pollinated seed from "Yellow Bell" segregated for normal red fruited (21 progeny) and yellow fruited (3 progeny), suggesting the trait is recessive. The three yellow-fruited progeny likely arose from self-fertilization.

The majority of fruit and vegetative traits appear to follow quantitative inheritance. The most important trait is yield and the consistency to yield year to year. Since fruit load consumes plant resources during the period when inflorescence buds are initiated for the following year's crop, biennial bearing is not an uncommon feature in cranberry production. The first generation hybrids developed by the USDA breeding program appeared to improve the year-to-year productivity of cranberry, by being more tolerant of environmental stresses, e.g., high nitrogen environment (Davenport and Vorsa 1999). Yield is a complex trait which represents the outcome of a number of vegetative and reproductive traits, which need to be in a "homeostatic" combination. Specific yield components include vegetative vigor, upright density, flower bud set, percent fruit set, fruit size, fruit set/upright, and gamete viability.

2.2.4 Breeding for New Varieties

The USDA, in cooperation with the New Jersey and Massachusetts State Agricultural Experiment Stations initiated the first cranberry breeding program in 1929 (Chandler et al. 1947). The major objectives of the program were to develop varieties resistant to "false-blossom", along with higher productivity and superior fruit. Over 10,685 seedlings were produced from over 30 crosses, and led to "the 40 selections" for further testing. The 40 selections were further evaluated for sauce and cocktail quality, specific gravity and overall appearance (Chandler et al. 1947). Traits evaluated included susceptibility to leafhopper feeding, date of harvest, size of fruit, decay, yield, and shape (Chandler et al. 1947). This program resulted in the release of seven varieties, including the most widely grown cultivar "Stevens" arising from a "McFarlin" x "Potter" cross which was field selected in 1940 at J.J. White Co., Whitesbog, Burlington Co., New Jersey and released in 1950 (Chandler et al. 1950). Other varieties released from this program include "Pilgrim", "Wilcox", "Franklin", "Bergman" and "Beckwith". In addition, selections not officially released and named also have been planted commercially, e.g., "No. 35". Washington State University also released the variety, "Crowley" in 1961, which was initially fairly widely planted but has been largely replaced.

Pollen is easily collected from flowers by holding the flower between the thumb and forefinger rolling and gently squeezing the flower. Pollen can be stored at 2–4°C for up to year or more. Pollen can be applied to stigmas 2–5 days post-emasculation. Seeds can be harvested once fruit is ripe, generally 1–2 weeks following color development. Seeds should be maintained in moist conditions at 1–4°C for 2–3 months, before sowing. Seeds should be sown in moist acidic media and seedlings can be transplanted to potted culture in peat/sand 1/1 v/v media. Irrigation should be with neutral to low pH water. Seedlings can be field planted directly, or propagated through cuttings to establish plots. Plots typically require 2–3 years to be fully colonized.

Cranberry requires a pollinator for pollination, and most commercial growers utilize honeybees at about 1–2 hives/acre. Cranberry pollination usually begins in early to mid June and finishes by early to mid July. The majority of fruit growth is completed by mid September. Fruit evaluation is initiated in late summer to identify early ripening progeny. Plots are typically rated for yield, fruit rot, vegetative cover, "runnering" (stolon production), upright density, vegetative diseases and fruit traits (fruit size, shape, etc.). Yield is estimated by usually harvesting fruit from square foot samples, where grams per square fruit translates approximately to barrels/ acre, the standard measure for cranberry production. Fruit traits include total anthocyanins (TAcy), percent soluble solids (Brix), and titratable acidity (citric acid equivalents). TAcy is measured in mg/100g fruit fresh weight by water extraction, filtration and absorbance at 520 nm (Vorsa et al. 2003).

2.3 Cranberry Genome

2.3.1 Cytology

Vaccinium species are possibly secondary polyploids with a primary basic number $x = 6$ (Ahokas 1971b). *V. macrocarpon* naturally exists as a diploid ($2n = 2x = 24$), while *V. oxycoccos* exists at the diploid, tetraploid ($2n = 4x = 48$) and hexaploid ($2n = 6x = 72$) levels (Ahokas 1971a, b). The genome is organized as 12 metacentric and submetacentric chromosomes ranging from 1.4 to 2.3 µ in length as observed during somatic metaphase (Hall and Galletta 1971).

Within *V. macrocarpon* pollen mother cells normally 12 bivalents, ring and rod bivalents, are observed during diakinesis in most varieties (Fig. 2-6). The cultivar "Howes" and its progeny "Wilcox" were identified as reciprocal translocation heterozygotes, exhibiting multivalents during pollen mother cell meiosis and about 50% pollen abortion (Ortiz and Vorsa 1998). Tetrad analysis of aborted pollen indicated significant environmental (year-to-year)

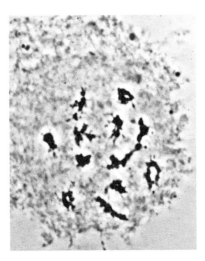

Figure 2-6 Diakinesis in cranberry.

variation for interstitial chiasma formation, as well as genetic background influence chiasma formation and recombination estimates (Ortiz and Vorsa 1998). Progeny from crosses with the translocation heterozygotes cultivars Howes and Wilcox, indicated presence of a semi-balanced lethal system located in the translocation segments (Ortiz and Vorsa 2004).

2.3.2 Markers

Sequence-characterized amplified region (SCAR) markers for cranberry germplasm analysis were developed using nine primer sets designed from RAPD-identified polymorphic markers (Polashock and Vorsa 2002). These primer sets generated 38 markers across a cranberry germplasm collection. They reported that RAPD and SCAR methods produced comparable results in 27 randomly chosen cranberry germplasm accessions and concluded that SCAR and RAPD markers can be employed for identifying closely related genotypes. However, SCARs provided more reproducible polymorphic markers than those of RAPDs.

Although RAPD and SCAR markers generated substantial polymorphism among cranberry germplasm, researchers used inter simple sequence repeat (ISSR) primers to distinguish wild lingonberry (*V. vitis-idaea* L.) clones and suggested that ISSR could be used for other *Vaccinium* species (Debnath 2007b). The ISSR primers target microsatellites that are abundant throughout the plant genome (Wang et al. 1994) and have proven to be more reproducible than RAPD markers, and generally reveal higher levels of polymorphism because of the higher annealing temperature and longer sequences of ISSR primers (Qian et al. 2001). They also cost less and are easier to use than amplified fragment-length polymorphisms (AFLPs) and do not require prior knowledge of flanking sequences like simple sequence repeats (SSRs) (Reddy et al. 2002). Researchers used expressed sequence tag (EST)-PCR markers for DNA-fingerprinting and mapping in blueberry that had previously been developed from ESTs produced from a cDNA library derived from RNA from floral buds of cold acclimated plants (Rowland et al. 2003). Because EST-PCR markers are derived from gene coding regions, they are more likely to be conserved across populations and species than markers derived from random regions of DNA, such as RAPD or AFLP markers. These markers may be useful for DNA-fingerprinting, mapping, and assessing genetic diversity within cranberry (Rowland et al. 2003). SSR markers were developed for blueberry using the same EST library and genomic sequences (Boches et al. 2005). A few of those developed for blueberry were successfully applied to cranberry (Bassil and Hummer 2007).

2.3.3 Genetic Transformation

Crop improvement through transformation can introduce new genes into cranberry that are unattainable through classical breeding methods. Direct transformation facilitates interspecies gene transfer and the potential to vastly improve pest and disease resistance, yield, nutritional content, and other desirable traits. Cranberry fruit rot presents considerable problems to growers and the introduction of antifungal genes could allow for reduced reliance on traditional fungicides. Similarly, the insertion and expression of genes for insect pest resistance into elite genotypes would offer farmers an alternative to traditional insecticides. Even without the introduction of interspecific genes, the manipulation of the existing cranberry genome is also an option, enhancing the native genes that control pest resistance and nutritional content.

In 1992 the first report of cranberry transformation was published. Serres and collaborators used electric discharge particle bombardment to successfully transform cranberry (Serres et al. 1992). The relatively low frequency (0.15%) of recovered transclones indicated that a more efficient DNA delivery system could be developed. Another disadvantage of particle bombardment is that chimeric plants can be created, and the insertion of multiple copies of the desired genes can lead to expression problems.

Transformation using *Agrobacterium tumefaciens* harboring disarmed Ti plasmids is a popular technique of delivering and incorporating DNA into plant cells. Wild-type *A. tumefaciens* infection (crown gall) has been observed in both blueberry and cranberry. Researchers investigating *A. tumefaciens* as a system for cranberry transformation have had little success in obtaining stable transclones (Polashock and Vorsa 2002).

Cranberry has shown an interesting interaction with the popular reporter GUS used to identify transclones. The high concentration of secondary plant compounds found in cranberry has been reported to inhibit GUS activity. Specifically the endogenous phenolic compounds in cranberry are the likely source for the unpredictable histochemical staining, as the addition of PVPP (which specifically absorbs phenolic compounds) somewhat reduced the inhibitory action (Serres et al. 1997). The use of another reporter gene such as green fluorescent protein (GFP) is recommended when transforming cranberry.

Genes with commercial interest are of primary importance in transgenic cranberry. Transgenic resistance to insects has been primarily explored through the inclusion of a gene from the bacterium *Bacillus thuringiensis* (*Bt*), which confers production of a sigma-endotoxin. The sigma-endotoxin is ingested by insects and cleaved into an insecticidal compound lethal to specific orders of insects. Although inhibition of sigma-endotoxin insecticidal activity due to high tannins has been reported in plant systems, it is unknown if the phenolic compounds of cranberry have an inhibitory effect. Transgenic *Bt* cranberry has shown disappointing results in deterring blackheaded fireworm, a significant agricultural pest, in field and laboratory tests.

Herbicide resistance genes could allow the use of broad spectrum herbicides without crop damage. Researchers at the University of Wisconsin-Madison have tested transgenic "Pilgrim" cranberry plants with expressing the *bar* gene which confers resistance to glufosinate-ammonium. Nontransformed plants were killed at 200–300 ppm and transformed plants survived 500 ppm glufosinate-ammonium. A second study using a more translocatable form of the herbicide, bialophos, was conducted. The transformed cranberry showed excellent resistance and only minor damage at the highest concentration tested (1600 ppm).

2.3.4 Genome Size

An estimate of *V. macrocarpon* genome size was conducted using flow cytometry using the grass *Brachypodium distachyon* (L.) P. Beauv. as a control. Compared to other fully sequenced plant genomes, the *V. macrocarpon* diploid genome is on the smaller side at 470 megabases (Table 2-2).

To get accurate cranberry genome size estimates, each genotype was run through flow cytometer three times and an average was generated from these estimates. The data from these experiments is summarized in Table 2-2. All *V. macrocarpon* samples were obtained from the New Jersey Agricultural Experiment Station in Chatsworth, New Jersey.

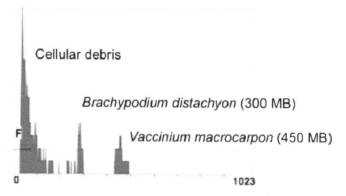

Figure 2-7 Flow cytometry measures the flourescence of propidium iodide stained nuclei of an internal reference standard (*Branchypodium distachyon*) and that of an unknown genome (*V. macrocarpon*). Intensity of flourescence is relative to genome size. The 2C nuclei peak of cranberry (*Vacciunium macrocarpon*) is 1.5×-fold larger than Brachypodium distachyon. Therefore, the *V. macrocarpon* haploid genome is estimated at 470 MB, based on the *B. distachyon* haploid genome size of 300 MB. The peak at the far left of the graph is cellular debris. Some variation in genome size within cranberry accessions was found (Table 2-3).

Table 2-3 Genome size varies among genotype in cranberry, estimates determined using flow cytometry.

Genotype	Avg. Genome Estimate (MB)
Stevens	465
NJ88-9-29	535
NJ99-109-2	458
NJ99-124-4	460
NJ98-310-4	451
NJ99-101-4	444

2.3.5 Why Sequence the Cranberry Genome?

Whole genome sequencing holds the potential to allow scientists a direct means of locating genes for the improvement of yield, disease resistance, and nutritional value of crop plants. The creation of highly inbred (homozygous) cranberry lines, the cytological investigation of cranberry chromosomes, and estimates of genome size, are early steps completed for the process of sequencing the cranberry genome. At approximately 470 megabases, the diploid genome of *V. macrocarpon* Aiton, the American cranberry, is an excellent candidate for whole genome sequencing. This niche crop is harvested as fresh fruit for juices, sauces, and other foods, and the fruit contains bioactive compounds that benefit human health. Current breeding programs seek to improve nutritional content, disease resistance, and pest deterrence. A sequenced genome combined with a physical map of cranberry traits would facilitate breeding efforts, provide growers superior varieties, and provide consumers an improved "super-food".

2.4 Future Research and Perspectives

Inbred lines have been developed for several cranberry accessions for use in high-resolution genetic mapping and marker identification. These markers will be developed from single nucleotide polymorphisms (SNPs) and SSRs and used to resolve linkage groups in cranberry. Marker-assisted selection of cranberry cultivars may speed the process of releasing new genotypes.

Sequencing and assembly of the cranberry genome, coupled with gene function and expression studies will provide a host of candidate genes for various practical functions in cranberry. A newly sequenced cranberry genome will provide raw data for study in disease resistance, cold acclimation, and nutritional quality. Cranberry genes involved in the pathways of bioactive compounds may contribute to the elucidation of the interaction of genes and gene products in fruit development, nutritional quality, and bioactivity. Comparison with *Vaccinium* genomes and other members of the family Ericaceae may provide new insights to cranberry phylogeny.

References

Ahokas H (1971a) Cytology of hexaploid cranberry with special reference to chromosomal fibres. Hereditas 68: 123–136.
Ahokas H (1971b) Notes on polyploidy and hybridity in *Vaccinium* species. Ann BotFenn 8: 254–256.
Aldrich P, Doebley J (1992) Restriction fragment variation in the nuclear and chloroplast genomes of cultivated and wild *Sorghum bicolor*. Theor Appl Genet 85: 293–302.
Avorn J, Monane M, Gurwitz JH, Glynn RJ, Choodnovskiy I, Lipsitz LA (1994) Reduction of bacteriuria and pyuria after ingestion of cranberry juice. JAMA 271: 751–754.
Bassil N, Hummer K (2007) Blueberry microsatellite markers identify cranberries. Int Soc for Hort Sci Acta Horticulturae. 810: 181–187.
Beachy E (1981) Bacterial Adherence: Adhesin-receptor interaction mediating the attachment of bacteria to mucosal surfaces. J Infect Dis 143: 325–345.
Boches P, Rowland L, Hummer K, Bassil N (2005) Microsatellite markers developed from 'Bluecrop' reveal polymorphisms in the genus *Vaccinium* and are suitable for cultivar fingerprinting. HortScience 40: 1122.
Boone D (1995) Vector unknown diseases: ringspot. In: F Caruso, D Ramsdell (eds) Compendium of Blueberry and Cranberry Diseases. Am Phytopathol Soc Press, St. Paul, MN, USA.
Bruederle L, Hugan M, Dignana J, Vorsa N (1996) Genetic variation in natural populations of the large cranberry *Vaccinium macrocarpon* Ait. (Ericaceae). Bull Torrey BotClub 123: 41–47.
Burger O, Ofek I, Tabak M, Weiss EI, Sharon N, Neeman I (2000) A high molecular mass constituent of cranberry juice inhibits *Helicobacter pylori* adhesion to human gastric mucus. FEMS Immunol Med Microbiol 29: 295–301.
Cane J, Schiffhauer D, Kervin L (1996) Pollination, foraging, and newsting ecology of the leaf-cutting bee *Megachile* (Delomedachile) *addenda* (Hymenoptera: Megachilidae) on cranberry beds. Ann Entomol Soc Am 89: 361–367.
Chandler F, Wilcox R, Bain H, Bergman H, Dermen H (1947) Cranberry breeding investigation of the U.S. Dept. of Agriculture. Cranberries 12: 6–9.
Chandler F, Bain H, Bergman H (1950) The Beckwith, the Stevens, and the Wilcox cranberry varieties. Cranberries 14: 6–7.
Chu YF, Liu RH (2005) Cranberries inhibit LDL oxidation and induce LDL receptor expression in hepatocytes. Life Sci 77: 1892–1901.
Dana MN (1983) Cranberry cultivar lista. Fruit Var J 37: 88–95.
Davenport J, Vorsa N (1999) Cultivar fruiting and vegetative response to nitrogen fertilizer in cranberry. J Am Soc Hort Sci 124: 90–93.
Debnath S (2005) Differentiation of *Vaccinium* cultivars and wild clones using RAPD markers. J Plant Biochem Biotechnol 14: 173–177.
Debnath S (2007a) An assessment of the genetic diversity within a collection of wild cranberry (*Vacinium macrocarpon* Ait.) clones with RAPD-PCR. Genet Resour Crop Evol 54: 509–517.
Debnath S (2007b) Inter simple sequence repeats (ISSR) to assess genetic diversity within a collection of wild lignonberry (*Vaccinium vitis-idaea* L.) clones. Can J Plant Sci 87: 337–344
Dolezel J, Bartos J (2005) Plant DNA flow cytometry and estimation of nuclear genome size. Ann Bot 95: 99–110.
Duarte S, Gregoire S, Singh AP, Vorsa N, Schaich K, Bowen WH, Koo H (2006) Inhibitory effects of cranberry polyphenols on formation and acidogenicity of *Streptococcus mutans* biofilms. FEMS Microbiol Lett 257: 50–56.
Ferguson PJ, Kurowska E, Freeman DJ, Chambers AF, Koropatnick DJ (2004) A flavonoid fraction from cranberry extract inhibits proliferation of human tumor cell lines. J Nutr 134: 1529–1535.

Foo LY, Lu Y, Howell AB, Vorsa N (2000a) A-Type proanthocyanidin trimers from cranberry that inhibit adherence of uropathogenic P-fimbriated *Escherichia coli*. J Nat Prod 63: 1225–1228.

Foo LY, Lu Y, Howell AB, Vorsa N (2000b) The structure of cranberry proanthocyanidins which inhibit adherence of uropathogenic P-fimbriated *Escherichia coli* in vitro. Phytochemistry 54: 173–181.

Gregoire S, Singh A, Vorsa N, Koo H (2007) Influence of cranberry phenolics on glucan synthesis by glucosyltransferases and *Streptococcus mutans* acidogenicity. J Appl Microbiol 103: 1960–1968.

Gu L, Kelm MA, Hammerstone JF, Beecher G, Holden J, Haytowitz D, Prior RL (2003) Screening of foods containing proanthocyanidins and their structural characterization using LC-MS/MS and thiolytic degradation. J Agri Food Chem 51: 7513–7521.

Guthrie N (2000) Effect of cranberry juice and products on human breast cancer cell growth FASEB J 2000 14: A771.

Hall S, Galletta G (1971) Comparative chromosome morphology of diploid *Vaccinium* species. J Am Soc Hort Sci 96: 289–292.

Harborne J (1997) Phytochemistry of fruits and vegetables: An ecological overview. In: F Thoman-Barberan (ed) Phytochemistry of Fruits and Vegetables. Oxford Univ Press, New York, USA, pp 353–367.

Howell AB, Vorsa N, Der Marderosian A, Foo LY (1998) Inhibition of the adherence of P-fimbriated *Escherichia coli* to uroepithelial-cell surfaces by proanthocyanidin extracts from cranberries. N Engl J Med 339: 1085–1086.

Jamieson A (2001) Horticulture in Canada—Spotlight on the Atlantic Provinces. Chron Hort 42: 8–11.

Kontiokari T, Sundqvist K, Nuutinen M, Pokka T, Koskela M, Uhari M (2001) Randomised trial of cranberry-lingonberry juice and *Lactobacillus* GG drink for the prevention of urinary tract infections in women. BMJ 322: 1571.

Kresty LA, Howell AB, Baird M (2008) Cranberry proanthocyanidins induce apoptosis and inhibit acid-induced proliferation of human esophageal adenocarcinoma cells. J Agri Food Chem 56: 676–680.

McKay DL, Blumberg JB (2007) Cranberries (*Vaccinium macrocarpon*) and cardiovascular disease risk factors. Nutr Rev 65: 490–502

McManus P, Best V, Voland R (1999) Infection of cranberry flowers by *Monilinia oxycocci* and evaluation of cultivar resistance to cotton ball. Phytopathology 89: 1127–1130.

McManus P, Caldwell R, Coland R, Best V, Clayton M (2003) Evaluation of sampling strategies for determining incidence of cranberry fruit rot and fruit rot fungi. Plant Dis 87: 585–590.

Neto C (2007) Cranberry and its phytochemicals: a review of in vitro anticancer studies. J Nutr 137: 186S–193S.

Ortiz R, Vorsa N (1998) Tetrad analysis with translocation heterozygotes in cranberry (*Vacinium macrocarpon* Ait.): Interstitial chiasma and directed chromosome segregation of centromeres. Hereditas 68: 75–84.

Ortiz R, Vorsa N (2004) Transmission of cyclical translocation in two cranberry cultivars. Hereditas 140: 81–86.

Oudemans P (1999) Detection, quantification, and identification of *Phytophthora* species associated with cranberry bogs. Plant Dis 83: 251–258.

Oudemans P, Caruso F, Stretch A (1999) Cranberry fruit rot in the northeast: a complex disease. Plant Dis 82: 1176–1184.

Oudemans P, Polashock J, Vinyard B (2008) Fairy ring disease of cranberry; assessment of crop losses and impact on cultivar genotype. Plant Dis 92: 612–622.

Penney B, Hendrickson P, Churchill R, Butt E (1997) The wild partridgeberry (*Vaccinium vitis-idaea* L. var. minus Lodd.) Industry in Newfoundland and Labrador and the potential for expansion utilizing European cultivars. Acta Hort 446: 139–142.

Polashock J, Vorsa N (2002) Development of SCARs for DNA fingerprinting and germplasm analysis of cranberry. J Am Soc Hort Sci 121: 210–215.

Polashock J, Vaicunis J, Oudemans P (2005) Identification of a new *Phytophthora* species causing root and runner rot of cranberry in New Jersey. Phytopathol News 95: 1237–1243.

Pozdnyakova L, Oudemans P, Hughes M, Gimenez D (2002) Spatial detection and quantification of *Phytophthora* root rot effects on cranberry yield. Comput Electron Agri 37: 57–70.

Qian Q, Ge S, Hong D (2001) Genetic variation within and among populations of a wild rice *Oryza sativa* from China detected by RAPD and ISSR markers. Theor Appl Genet 102: 440–449.

Reddy P, Sarla N, Siddiq E (2002) Inter simple sequence repeat (ISSR) polymorphism and its application in plant breeding. Euphytica 128: 1428–1432.

Robideau G, Caruso F, Oudemans P, McManus P, Renaud M, Auclair M, DeVarna J, Levesque C (2008) Detection of cranberry fruit rot using DNA array hybridization. Can J Plant Pathol 30: 226–240.

Rowland L, Dhanaraj A, Polashock J, Arora R (2003) Utility of blueberry-derived EST PCR primers in related Ericaceae species. HortScience 38: 1428–1432.

Ruel G, Couillard C (2007) Evidences of the cardioprotective potential of fruits: the case of cranberries. Mol Nutr Food Res 51: 692–701.

Sarracino J, Vorsa N (1991) Self and cross fertility in cranberry. Euphytica 58: 129–136.

Serres R, Stang E, McCabe D, Russell D, Mahr D, McCown B (1992) Gene transfer using electric discharge particle bombardment and recovery of transformed cranberry plants. J Am Soc Hort Sci 117: 174–180.

Serres R, McCown B, Zeldin E (1997) Detectable beta-glucuronidase activity in transgenic cranberry is affected by endogenous inhibitors and plant development. Plant Cell Rep 16: 641–646.

Steinberg D, Feldman M, Ofek I, Weiss EI (2004) Effect of a high-molecular-weight component of cranberry on constituents of dental biofilm. J Antimicrob Chemother 54: 86–89.

Stewart CJ, Nilsen E (1995) Phenotypic plasticity and genetic variation of *Vaccinium macrocarpon*, the American cranberry. II. Reaction norms and spatial clonal patterns in two marginal populations. Int J Plant Sc 156: 698–708.

Stiles C, Oudemans P (1999) Distributions of cranberry fruit-rotting fungi in New Jersey and evidence for host nonspecific resistance. Phytopathology 89: 218–225.

Teifon Jones A, McGavin W, Dolan A (2001) Detection of tobacco streak virus isolates in North American cranberry (*Vaccinium macrocarpon* Ait.). Plant Health Progr Online, pp doi: 10.1094/PHP-2001-0717-1001-RS.

Trehane J (2004) Blueberries, Cranberries, and other *Vacciniums*. Timber Press, Inc, Portland, USA.

USDA (2008) National Agricultural Statistics Service. USDA.

Vander Kloet S (1988) The Genus *Vaccinium* in North America. Can Govt Publ Center, Ottawa, Canada.

Vasanthakumar A, McManus P (2004) Indole-3-acetic acid-producing bacteria are associated with cranberry stem gall. Phytopathology 94: 1164–1171.

Vorsa N, Polashock J, Cunningham D, Roderick R (2003) Genetic inferences and breeding implications from analysis of cranberry germplasm anthocyanin profiles. J Am Soc Hort Sci 128: 691–697.

Vvedenskaya IO, Rosen RT, Guido JE, Russell DJ, Mills KA, Vorsa N (2004) Characterization of flavonols in cranberry (*Vaccinium macrocarpon*) powder. J Agri Food Chem 52: 188–195.

Walker EB, Barney DP, Mickelsen JN, Walton RJ, Mickelsen RA, Jr. (1997) Cranberry concentrate: UTI prophylaxis. J Fam Pract 45: 167–168.

Wang Z, J W, Zhong G, Tanksley S (1994) Survey of plant short random DNA repeats. Theor Appl Genet 88: 1–6.

Wegener L, Punja Z, Martin R (2004) First report of blueberry scorch virus in cranberry in Canada and the United States. Plant Dis 88: 427.

Weiss EL, Lev-Dor R, Sharon N, Ofek I (2002) Inhibitory effect of a high-molecular-weight constituent of cranberry on adhesion of oral bacteria. Crit Rev Food Sci Nutr 42: 285–292.

Zhang L, Ma J, Pan K, Go VL, Chen J, You WC (2005) Efficacy of cranberry juice on *Helicobacter pylori* infection: a double-blind, randomized placebo-controlled trial. Helicobacter 10: 139–145.

3

Blackberries and Raspberries

John-David Swanson,[1] John E. Carlson,[2] Felicidad Fernández-Fernández,[3] Chad E. Finn,[4] Julie Graham,[5] Courtney Weber,[6] and Daniel J. Sargent[3]

3.1 *Rubus*: A Preamble

3.1.1 *Economic Importance, Production Areas, and Nutritional Composition*

Blackberries (*Rubus* sp. L.), red raspberries (*R. idaeus* L. and *R. strigosus* Michx) and black raspberries (*R. occidentalis* L.) are grown in many areas of the world, but they are most productive in regions with mild winters and moderate summers. While generally referred to as "brambles" in the eastern US, they are generally known as "caneberries" in the western US. The European red raspberry (*R. idaeus*) and the American red raspberry (*R. strigosus*) are often combined by taxonomists into a single *R. idaeus* genus or considered separate subspecies of the same genus—*R. idaeus* subsp. *idaeus* and *R. idaeus* subsp. *strigosus*.

Blackberries and red raspberries are sold fresh, primarily in clam shell packages and sometimes as processed products, whereas the bulk of the black raspberry crop is processed. The primary processed caneberry

[1]Department of Biology, University of Central Arkansas, 201 Donaghey Ave., Conway, AR 72032, USA.
[2]The School of Forest Resources, Department of Horticulture, and the Huck Institutes of the Life Sciences, Pennsylvania State University, 323 Forest Resources Building, University Park, PA 16802, USA.
[3]East Malling Research, New Road, East Malling, Kent, ME19 6BJ, UK.
[4]US Dept of Agriculture- Agricultural Research, Service, Horticultural Crops Research Lab, 3420 NW Orchard Ave., Corvallis, OR 97330, USA.
[5]SCRI, Invergowrie, Dundee, DD2 5DA, UK.
[6]Department of Horticultural Sciences, Cornell University, New York State Agricultural Experiment Station, Geneva, NY 14456, USA.

products include fruit that is individually quick frozen (IQF), bulk frozen as whole fruit, pureed, juiced, canned, or dried. From these fundamental wholesale products, a host of products are made for the retail market and for institutional food service products.

Blackberry production is rapidly increasing (Strik 1992; Clark 2005; Clark et al. 2007; Strik et al. 2007), and there were an estimated 140,292 MT commercially harvested from 20,035 ha in 2005. Europe leads the world in acreages harvested (7,692 ha), while North America has the greatest production (59,123 MT). Serbia (69%) dominates European production, but a number of countries have significant production. In North America, Oregon is the major producer. However, Mexican production has been rapidly increasing, particularly in Michoacan and Jalisco in the past five years. California and Arkansas are the only other states in the US with over 1,000 MT in production. Central American production (1,620 ha) is predominantly from Costa Rica and Guatemala where, in addition to harvest from managed stands, a great deal is harvested from feral stands. South American production (1,597 ha) is predominantly from Ecuador and Chile. Asian production has been rapidly increasing with over 1,550 ha of new plantings predominantly in China. Jiangsu is the most productive province, although Liaoning, Shandong and Hubei are increasing their production. The production in Oceania is mainly in New Zealand although the area planted is small with only about 259 ha. African production is only reported in South Africa but has been initiated in Morocco, Algeria, Kenya and possibly others. The bulk of the fruit is grown for processing applications in the Pacific Northwest US, Serbia, and China, whereas fresh market sales are the focus of the industry elsewhere.

The major production areas of red raspberries in North America are the Pacific Northwest (Oregon, Washington and British Columbia), California, the eastern US (New York, Michigan, Pennsylvania and Ohio) and a rapidly expanding industry in Mexico. In Europe, red raspberries are grown to the largest extent in Serbia, Russia and Poland, with commercial production scattered all across the European Union. Fresh market production is expanding rapidly in Spain, Portugal and other regions with a climate suitable to mimic the California production system. Average world production from 2005–2007 was estimated at 602,500 MT from an area of 71,250 ha (wild plants were not included except from the Russian Federation) (Hall et al. 2009).

While adapted to many of the same areas as the other cultivated *Rubus*, black raspberry (*R. occidentalis*) cultivation is concentrated in Oregon in the western US, Ohio, Pennsylvania, and New York in the eastern US and in Korea. Oregon has traditionally been the leading producer with about 600 ha producing ~ 5,200 MT almost all for processing. The eastern US production is estimated to be around 100 ha, and the fruit is grown for fresh and processing markets. While black raspberry production in Korea is estimated to be around 1,332 ha and 5,500 MT (S J Yun, pers. comm.), this figure may include other *Rubus* sp. and reflects a three-fold increase in acreage since 2003; most of the Korean production is used to make a liqueur called Bokbunja.

In addition to being rich in the traditionally evaluated nutrients (Finn 2008b; Moore et al. 2008), the caneberries have among the highest levels of antioxidants/phytonutrients of any fruit crop due primarily to their intense concentration of anthocyanins and phenolic compounds (Moyer et al. 2002). A considerable amount of new research has been performed on variation patterns in the antioxidant capacity of *Rubus* species and crosses. The fact that anthocyanins and polyphenolics are powerful antioxidants has led to a number of investigations that have studied the nutraceutical/antioxidant levels of raspberries and will be discussed in detail later (Moyer et al. 2002; Perkins-Veazie and Kalt 2002; Wada and Ou 2002; Siriwoharn et al. 2004; Anttonen and Karjalainen 2005; Beekwilder et al. 2005; Dossett et al. 2008; Moore et al. 2008; Weber et al. 2008).

3.1.2 Academic Importance: Use as a Model Plant in Genetics, Cytogenetics, Breeding and Genomics

Biotechnology has resulted in a fundamental shift in detecting and monitoring genetic variation in plant breeding and genetic studies. A variety of molecular marker techniques including isozymes, random amplified polymorphic DNA (RAPD), simple sequence repeats (SSR), amplified fragment length polymorphism (AFLP) and others, have been employed in genetic studies of raspberry and blackberry. *Rubus* species have been utilized for the construction of genetic linkage maps and the study of specific mechanisms of pest resistance. Wild *Rubus* populations have been investigated to elucidate diversity and population dynamics (Graham and McNicol 1995; Patamsytė et al. 2004). Additionally they have been used to assess gene flow in wild populations as a model for ecological effects of habitat fragmentation (Graham et al. 1997, 2003) and gene flow between wild populations and cultivars. The diversity of the weedy raspberry species *R. alceifolius* Poir. has been investigated in its native range and in areas where it has invaded as a noxious weed (Amsellem et al. 2001a).

A hydroponic system was developed to directly observe the progression of disease to elucidate root rot in perennial species. Generational means analysis was then combined with molecular markers and quantitative trait loci (QTL) analysis to map resistance to Phytophthora root rot (*Phytophthora fragariae* var. *rubi* Hickman) in a red raspberry population (Pattison et al. 2007). Attempts to develop markers for viral resistance genes have been carried out for raspberry leaf spot and raspberry vein chlorosis (Rusu et al. 2006) utilizing the "Glen Moy" × "Latham" cross of Graham et al. (2004).

Additionally, many genomics tools are becoming available (discussed in Chapter 3.8). These include the construction of the first publicly available red raspberry bacterial artificial chromosome (BAC) library from "Glen Moy". Currently, the library comprises over 15,000 clones with an average insert size of approximately 130 kb (6–7 genome equivalents). Additionally, there has been several gene expression libraries created from various organs of the plant. Together, these libraries are allowing researchers to identify genes and elucidate their function at a much increased rate compared to the past.

3.1.3 Brief History of the Crop: Center of Origin, Botanical Origin and Evolution, Domestication, and Dissemination

The center of diversity is considered to be in China where there are 250–700 species of *Rubus* depending on the taxonomist (Thompson 1997). While species are found on all continents except Antarctica, the greatest number of species is found in Eurasia and North America.

Most *Rubus* species were likely food sources wherever humans found them. *Rubus* has been traced back through ancient and historical times through artwork or illustrations (Hummer and Janick 2007). European blackberry and red raspberry plants were mentioned by Ancient Greeks and Romans. At Newberry Crater near Bend, Oregon (USA), artifacts of food remnants containing *Rubus* date to 8,000 BCE. The writings of Aeschylus and Hippocrates from 500–400 BCE include caneberries. By 370 BCE, ancient Greeks were harvesting raspberries. Pompey brought raspberries from southeast of Troy, in current day Turkey, to the Romans around 65 BCE. The Hebrew Bible contained many references to thorny plants native to the Holy Land that some have attributed to *R. sanctus* Schreb. or *R. ulmifolius* Schott (Hummer and Janick 2007). The term *sĕneh* used to describe these species is also the term used in Exodus 3:1–5 to describe God's appearance to Moses "in the flame of fire in the bush" (Hummer and Janick 2007). Numerous herbals, particularly Dioscorides' *De Materia Medica* written about 65 CE included the curative properties of blackberry or raspberry (Hummer and Janick 2007). The *Juliana Anicia Codex* from around 512 CE has the first image of *Rubus* that survived antiquity. During the Renaissance, *Rubus* was well represented and Hummer and Janick (2007) cite two paintings by Jan Bourdichon illustrating *Horae ad isum Romanum*, a prayer book of Anne of Bretagne (1503–1508), a drawing of Leonardo da Vinci (1510–1512) and a woodcut from *De Historia Stirpium* (1544), and a herbal by Leonhart Fuchs, as particularly good examples (Online versions of Fuchs work and others may be found at the Cushing/Whitney Historical Medical Library at Yale University (*http://www.med.yale.edu/library/historical/fuchs/#50*).

Raspberries gradually grew in popularity over the centuries and by the 1500s, *R. idaeus* was cultivated all over Europe. In 1829, 23 cultivars were listed in the "History of English Gardening" (Finn and Hancock 2008). North American *R. strigosus* was introduced into Europe in the early 19th century and natural hybrids with *R. idaeus* resulted in many improvements. Most cultivars dating from this period are hybrids of these two species (Daubeny 1983; Dale et al. 1989, 1993).

By the 1600s, blackberries were being mentioned in gardening books (Jennings 1988). Since blackberries were so common in civilization there seemed to be little interest in domestication and identification of superior genotypes, let alone breeding, until the 1800s. Not surprisingly, some of the first recorded selections from the wild were oddities such as albino or pink selections (Hedrick 1925).

3.1.4 Brief History of Breeding

Several excellent reviews of blackberry and raspberry breeding have been written in the past couple of years including Clark et al. (2007), Finn (2008a) and Finn and Hancock (2008).

3.1.4.1 Blackberry: Logan, Young, Boysen, and a Popular Berry Farm

Judge James H. Logan of Santa Cruz, CA (USA) is usually credited with making the first blackberry breeding effort. "Loganberry", released in 1890, and "Black Logan" were the most successful from his program; although, "Loganberry", which is still grown commercially more than a century after its release, was selected from open-pollinated fruit of the pistillate "Aughinbaugh" presumably crossed with "Red Antwerp" red raspberry (Logan 1955). Luther Burbank who, along with Thomas Edison and Henry Ford (played such a powerful role in the public's imagination), was active in blackberry breeding as well and developed "Phenomenal"/ "Burbank's Logan" that was nearly indistinguishable from "Loganberry" (Darrow 1925; Clark et al. 2007). Byrnes M. Young, an amateur horticulturist in Morgan City, LA (USA) could not grow "Loganberry" or "Phenomenal" but following correspondences with Burbank, decided to make crosses in 1905 between "Phenomenal" and the locally adapted "Austin Mayes" to produce "Youngberry". "Youngberry" was released in the 1920s, and is still commercially grown today. "Youngberry" has a prominent place in pedigrees as it is a grandparent of the widely grown "Marion" (Clark et al. 2007). The origin of the trailing blackberry "Boysenberry" from the same era is unknown, but Wood et al. (1999) presented a thorough examination of its history. "Boysenberry" was discovered by Rudolph Boysen on the Lubben farm in Napa County in northern California (USA). Later when he moved to southern California, the plants caught the attention of George Darrow, a plant breeder from the United States Department of Agriculture—Agricultural Research Service (USDA-ARS) (Beltsville, MD) who convinced a local grower and nurseryman, Walter Knott, to trial this selection. Knott and Darrow released it as a cultivar bearing the discoverer's name. Knott went on to develop a thriving business that started as a farm stand serving "Boysenberry" pie and has now become the Knott's Berry Farm entertainment and food products empire (Wood et al. 1999). While Wood et al.'s (1999) explanation of the historical origins of "Boysenberry" is well researched; there is still no certainty on the genetic origins of "Boysenberry". Similar hybrid berries between raspberry and blackberry such as "Laxtonberry", "Veitchberry", "Mahdi", and "Kings Acre" were developed in Europe (Darrow 1937).

"Thornless Evergreen", a selection of *R. laciniatus* Willd., is the final blackberry cultivar that was selected from the wild and is still commercially viable and is central to modern germplasm development (Waldo 1977). While the original cultivar "Evergreen" can be traced to the 1800s in Europe, the thornless selection later named "Thornless Evergreen" was found in Stayton, Oregon in 1926 and quickly became the industry standard. The thornless chimeral form is unstable and commonly reverts to thorny canes with environmental or mechanical injury. "Everthornless", which is genetically thornless, was developed from somaclonal plants developed from the L_1 layer of "Thornless Evergreen" (McPheeters and Skirvin 1983; McPheeters and Skirvin 1989, 2000).

Beginning in the early 1900s, public breeders were supported with public funding via the land grant university systems or the USDA-ARS. The Texas Agricultural Experiment Station breeding program was the first blackberry program (Darrow 1937). The primary goals of this program were to develop "hybrid berries" that were adapted to hot climates with low chilling. "Nessberry", developed there using *R. trivialis* Michx. germplasm, had some popularity, but it was even more valuable as a parent of "Brazos", which until recently was the most popular Mexican cultivar.

The John Innes Horticultural Institute in England and the New York State Agricultural Experiment Station followed by the USDA-ARS in Georgia were the next to develop programs. The John Innes program developed the critically important "Merton Thornless", which is the primary source of thornlessness in all tetraploid cultivars. The New York program developed several erect cultivars in the 1950s including "Bailey", "Hedrick", and "Darrow".

The major concerted breeding efforts worldwide that are still active are the University of Arkansas, the USDA-ARS in Oregon and the private program run by Driscoll's Strawberry Associates (Watsonville, CA) (Finn and Knight 2002). Two significant programs that have been discontinued or lost significant funding recently are the USDA-ARS program in Beltsville, MD and the Plant and Food Research Institute of New Zealand (previously HortResearch) program.

The USDA-ARS Beltsville program incorporated thornlessness from "Merton Thornless" into the first outstanding thornless cultivars released in the late 1960s and early 1970s ("Black Satin", "Smoothstem", "Thornfree", and "Dirksen Thornless") (Scott et al. 1957; Moore 1997). The USDA-ARS had a significant effort at their station in Carbondale, IL in the 1960s until it was closed in the early 1970s and "Hull Thornless" and "Chester Thornless" came out of this effort. The last release from these eastern USDA-ARS programs was "Triple Crown" in the 1990s (Galletta et al. 1998). This group of breeding material and cultivars is called "semi-erect" and the plants are characterized as being thornless, with very vigorous, large erect canes that grow 4–6 m long from a crown and arch to the ground. Their fruit is similar in quality to the erect blackberries and they are often incredibly productive.

The Plant and Food Research Institute of New Zealand, Ltd. Program (at the time HortResearch) was one of the most valuable and aggressive programs in the 1980s and 1990s as it developed "Boysenberry-like" cultivars and developed the "Lincoln Logan" source of spinelessness (S_{fl}) (Hall et al. 1986a, b c, 2002). "Ranui", "Waimate", "Karaka Black", and "Marahau" have been the notable recent releases from this program (Hall and Stephens 1999; Clark and Finn 2002; Hall et al. 2003).

Started in 1928, the USDA-ARS program in Oregon is the oldest continuously active blackberry breeding program. The efforts there combined wild selections (e.g., "Zielinski") of the Pacific Coast native, trailing, dioecious *R. ursinus* Cham et. Schlt. with a perfect flowered gene pool including "Loganberry", "Youngberry", "Himalaya", "Santiam"/ "Ideal", and "Mammoth", and cultivars from elsewhere, were used to develop cultivars for a whole new industry based on trailing blackberries. These crown-forming types that trail on the ground until lifted onto a trellis have excellent fruit quality but often have poorer winter hardiness than the other types. "Pacific", "Cascade", "Chehalem", and "Olallie" released from 1942–1950 were instrumental in the industry's establishment (Waldo and Wiegand 1942; Waldo 1948, 1950). "Marion", released in 1956, later became the industry standard in the Pacific Northwest (Waldo 1957). During the 1970s and 1980s this program worked to improve the fruit quality in the thornless germplasm pool that had been derived with the "Austin Thornless" source of thornlessness. The first trailing thornless cultivar to be released and commercially grown was "Waldo" (Lawrence 1989). Recent releases from this program are being widely planted in the Pacific Northwest, coastal California and other mild climate areas in Europe and South America; these include the thornless "Black Diamond",

"Black Pearl", and "Nightfall" as well as the very early ripening "Obsidian" (Finn et al. 2005a, b, c, d).

In addition to the trailing and semi-erect blackberries, the third type of blackberry is the erect blackberry, a selection that was largely developed by the University of Arkansas from eastern North American blackberry species. These types produce 1–4 m tall stiff upright canes and the plants sucker to produce a hedgerow. While there are erect cultivars, such as "Eldorado", which can be traced back to the 1800s, these really were developed as a viable commercial crop by the University of Arkansas beginning in 1964. The erect blackberry germplasm pool is similar to that of the semi-erects as they are all tetraploid and have comparable fruit quality characteristics. As with the semi-erects, thornlessness from "Merton Thornless" was incorporated into this gene pool. In 1989, "Navaho" was the first thornless erect cultivar to be released. Other cultivars released from this program include "Cheyenne" and "Cherokee" released in the 1970s, "Shawnee" in the 1980s, "Kiowa", "Apache" and "Chickasaw" in the 1990s, and "Ouachita" and "Natchez" in the 2000s. The Arkansas program, with some input from North Carolina State University, developed primocane fruiting blackberries. Brambles are generally biennial producing primocanes in the first season, and flowering floricanes the following year. Primocane fruiting blackberries are able to bypass the two year flowering cycle and produce flowers on primocanes. The plants are cut to the ground each year and then flower and fruit late in the season on new growth. "Prime-Jan" and "Prime-Jim" are the first of this type. Primocane fruiting was critical to the worldwide expansion of the red raspberry industry, and it is hoped that it will have a similar impact on blackberry.

Smaller sized and very productive programs are active elsewhere as demonstrated by the recent development of "Tupy" from Brazil (Clark and Finn 2002), "Loch Maree", "Loch Ness", and "Loch Tay" from the Scottish Crop Research Institute (SCRI) (Jennings and Brydon 1989; Clark and Finn 2008), "Chesapeake: from the University of Maryland (Clark and Finn 2002), and "Cacanska Bestrna" ("Cacak Thornless") from Serbia (Clark and Finn 1999; Stanisavljevic 1999).

In general, breeding programs are currently focusing on combining traits associated with fruit quality (esp. sweetness, color retention, and fruit firmness) with primocane flowering, thornlessness, erect canes, and heat tolerance (JR Clark, G Fernandez, F Fernandez-Fernandez, and C Finn, pers. comm.). However, other traits are also being investigated including plant productivity, plant health and vigor, cold tolerance, double blossom resistance, the ability to use mechanized harvesters, and virus tolerance (Table 3-1; Fig. 3-1).

Table 3-1 Top traits currently being investigated by raspberry and blackberry breeding programs (listed in perceived order of importance). *Represents fruit quality traits that were singled out separately as important.

Blackberry	Raspberry
Fruit quality	Yield
Sweetness*, intense color and flavor, color retention*, firmness*, seedlessness, high Brix, high titratable acidity, pH<3.5, uniform shape)	Phytopthora root rot and rust resistance
	Primocane fruiting
Primocane fruiting	Fruit quality
Thornlessness	*(Fruit firmness*, sweetness*, size*, size*, shelf life*,intense color and flavor, high titratable acidity, pH<3.5, uniform fruit shape)*
Erect canes	
Heat tolerance	Heat tolerance
Plant productivity	Insect resistance (esp. for virus vectors)
Cold tolerance in FF types	Virus tolerance
Virus tolerance	Cane erectness
Double blossom/rosette tolerance	Cold tolerance
Plant health/vigor	Thornlessness
Machine harvestable	Machine harvestable

Note: Compiled with the assistance of J. R. Clark, G. Fernandez, F. Fernandez-Fernandez, and C. Finn.

3.1.4.2 Raspberry

Finn and Hancock (2008) along with Jennings (1988) give good overviews of red raspberry breeding and much of the discussion below is a synopsis of their overviews. The first formal breeding work on raspberries began in North America. Dr. Brinkle of Philadelphia, PA was cited by Darrow (1937) as the "first successful raspberry breeder of this country" (Darrow 1937). Introduced by the Minnesota Breeding Farm in 1914, "Latham" has been the most enduring cultivar from this early breeding period and is still grown. "Pruessen", "Cuthbert" and "Newburgh" were developed in Europe and are hybrids between the North American and European species. These cultivars along with "Lloyd George" and "Pyne's Royal" (which are pure *R. idaeus*) are all cultivars that played an integral role in early red raspberry breeding.

"Lloyd George" was in the direct ancestry of 32% of the North American and European cultivars in 1970 (Oydvin 1970). This cultivar contributed several important traits including primocane fruiting, large fruit size and resistance to the American aphid. Jennings (1988) speculates that the success of "Lloyd George" hybrids "was possibly achieved because they combined the long-conical shape of "Lloyd George" receptacle with the

Figure 3-1 Common traits of interest found in blackberry. A) A terminal inflorescence of a primocane-fruiting genotype in summer, with a floricane fruit picked from the same plant. B) Long fruit of "Natchez" blackberry, a shape desired by marketers and growers. C) Poor fruit set of a primocane-fruiting genotype due to high heat during summer bloom. D) Fruit set of Prime-Jim in moderate climate during bloom in coastal California. E) Red drupe development after harvest and storage; preference is for retention of full black color. F) White drupe development, thought to be due to effects of intense ultraviolet light. G) Stem of the thorned genotype "Prime-Jim" F) Stem of the thornless genotype "Arapaho". (Photos A–F kindly provided by J. R. Clark, G-H provided by J. D. Swanson).

Color image of this figure appears in the color plate section at the end of the book.

more rounded shape of the American raspberries". "Willamette" from a cross of "Newburgh" × "Lloyd George" is an example of a "Lloyd George' hybrid" that dominated the industry in western North America for over a half century.

While many programs have released very successful cultivars in the 20th and early 21st centuries, a few programs have been particularly productive in releasing new germplasm. The East Malling program in the UK was responsible for the "Malling series". A number of selections were made prior to World War II and released in the 1950s including, "Malling Promise", "Malling Exploit" and the most successful, "Malling Jewel" (Jennings 1988). This program continues to have an impact with the recent release of "Octavia" (Finn et al. 2007). In addition to these floricane cultivars, the program has developed a number of very important primocane fruiting cultivars in the "Autumn series".

The very important "Glen series" was developed at the Scottish Crop Research Institute (UK). Their first release was "Glen Clova" in 1969, but the release of "Glen Moy" and "Glen Prosen" in 1981 offered great improvements in fruit size and flavor along with spinelessness. "Glen Ample", released in 1994, is one of the standards in the European wholesale raspberry market.

The breeding programs in the Pacific Northwest of North America at Washington State University (WSU; Puyallup, WA), Agriculture and Agri-Foods Canada (AAFC; Agassiz, BC) and the USDA-ARS (Corvallis, OR) have worked closely for decades. Together with programs in the UK they have made fantastic gains. The USDA-ARS's releases from the mid 1900s, "Willamette" and "Canby", are still commercially important floricane cultivars and their recent release "Coho" has been widely planted for its high yields of individual quick frozen (IQF) fruit (Finn et al. 2001). "Summit" and "Amity" primocane fruiting types from this program have been important since their release. "Summit" has seen a fantastic resurgence in interest due to its adaptation to Mexican production systems. "Meeker", developed by WSU and released in the 1960s, is still the processing industry standard (Finn 2006). This program also continues to be active and the newest releases "Cascade Delight" and "Cascade Bounty", which are root rot tolerant, and "Cascade Dawn" are being widely planted (Moore 2004, 2006; Moore and Finn 2007).

The AAFC program has played a primary role in developing extremely high quality fresh market cultivars over the past few decades. They took full advantage of germplasm exchanges with the UK and were successful at identifying outstanding selections out of crosses between British Columbia selections and some of the "Glen series", particularly "Glen Prosen" (Finn 2006). Their early releases such as "Chilcotin", "Skeena" and "Nootka" had excellent fruit quality and high yields for a fresh market berry. The program followed these releases with "Chilliwack" in the mid 1980s and "Tulameen" in 1989, which has become the world standard for fresh market fruit flavor and quality. This program remains active and the recent releases "Chemainus", "Saanich" and "Nanoose" are being widely planted (Kempler et al. 2006, 2007).

In the eastern US, Cornell University's New York State Agricultural Experiment Station (Geneva) has the oldest continuous raspberry breeding program in North America, dating to the late 1800s. Floricane cultivars such as "Taylor" and "Hilton" were early staples of the eastern industry. These have been replaced in recent decades with cultivars such as the large fruited "Titan", released in 1985, early season "Prelude" and very late season "Encore", released in 1998. "Titan" has proven to be an excellent parent, producing large fruited offspring, but it is susceptible to Phytophthora root rot.

In the 1950s primocane fruiting germplasm within the program in combination with material such as "Durham" (developed in New Hampshire) was used to produce an excellent primocane fruiting germplasm pool that culminated with the release of "Heritage" in 1969 and "Ruby" ("Watson") in 1988 (Daubeny 1997). First viewed as a novelty, the primocane fruiting types have revolutionized raspberry production. They have become the standard in regions where cold winter temperatures caused considerable winter damage to canes of floricane fruiting raspberries, as well as in low chill areas where floricane cultivars do not receive adequate chilling to be productive. Private companies in California, such as Driscoll's Strawberry Associates (Watsonville, CA), have developed cultivars and whole new production systems based around these primocane fruiting types where the plants were only in the ground for 18 months (Finn and Knight 2002). These cultivars and management systems have led to the rapid expansion of the California raspberry industry as well as industries in southern Europe, northern Africa, Australia and South Africa.

While the University of Minnesota program has been discontinued, their releases "Latham" and "Chief" were valuable commercial floricane cultivars and very valuable parents in further breeding. The University of Maryland has coordinated a breeding program with Virginia Tech University, Rutgers University, and the University of Wisconsin-River Falls, and the primocane fruiting "Caroline", "Anne", and "Josephine" developed in this program have become standards throughout most of North America.

The eastern North American black raspberry (*R. occidentalis*) was not cultivated until the 19th century, probably because of its abundance in the wild and the public's preference for red raspberry (Jennings 1988). Other than at Cornell's New York State Agricultural Experiment Station, there has been no consistent long term breeding effort in black raspberry. The station initiated its efforts in the late 1800s and was the primary center of research for much of the 20th century (Slate 1934; Slate and Klein 1952; Ourecky and Slate 1966; Ourecky 1975). However, by the late 20th century, there was little breeding effort at Geneva or anywhere else; only three cultivars were released in the last half of that century. There has been a recent resurgence in black raspberry interest at the New York State Agriculture Experiment Station, the USDA-ARS in Oregon and in Maryland, and with Plant and Food Research of New Zealand Ltd. In Oregon, Dossett et al. (2008) evaluated black raspberry families from sibling families from crosses among cultivars and a North Carolina selection to assess variation and inheritance of vegetative, reproductive and fruit chemistry traits in black raspberry. In New Zealand, "Ebony", the first spineless black raspberry cultivar, was released (H K Hall, pers. comm.), while in Maryland, an expressed sequence tag (EST) library was being fine tuned for use in black raspberry (K Lewers, pers. comm.), and finally in New York, a large scale breeding effort was initiated.

Driscoll Strawberry Associates, Inc. (Watsonville, CA), a private breeding program, has been breeding red raspberries in same manner since the 1930s. The program was reinitiated in a structured way in the early 1980s and their blackberry program started in 1991. While it may be irrelevant to others what Driscoll's does as the cultivars they develop are kept within the company, they have played an important role in the expansion of the fresh raspberry and blackberry industry. Additionally, the acreage devoted to their cultivars in North America and increasingly elsewhere is so large, that it is important to recognize their successful efforts.

Current breeding efforts in raspberry are focusing primarily on yield, RBDV and Phytophthora resistance, primocane fruiting, and fruit quality traits (including fruit firmness, sweetness, fruit size and fruit shelf life) (PP Moore, C Kempler, G Fernandez, F Fernandez-Fernandez, and C Finn, pers. comm.). Additional traits are also being investigated and include: heat and cold tolerance, insect and virus resistance, cane erectness, thornlessness, and the ability to use mechanized harvesters (Table 3-1).

3.1.5 Botanical Features

Rubus is a diverse group of hundreds of species divided botanically among 15 subgenera, many of which have been used in breeding (Jennings et al. 1991; Knight 1993; Finn 2001; Finn 2002a; Finn 2002b). The subgenus *Rubus* is divided into 12 sections with most of the cultivated blackberries being derived from the *Allegheniensis, Arguti, Flagellares, Rubus, Ursini,* and/or *Verotriviales* (Finn 2008a). Red and black raspberries along with many of the wild harvested species from around the world are in the *Idaeobatus* subgenus. The raspberries and blackberries are botanically separated by whether the torus (receptacle) remains in the fruit when picked in which case it is considered a blackberry, or remains on the plant leaving a hollow center to the fruit, in which case it is considered a raspberry. The separation leads to some incongruities when something like a "Loganberry", which has the shape and color of a red raspberry, is by definition a blackberry because the torus picks with the fruit.

The caneberries are typically found as early colonizers of disturbed sites and along forest edges. They are attractive to frugivores which move them quickly into open areas. As a group they are generally found in habitats with decent moisture levels. While they can be found in desert environments, they are always found in close proximity to springs, seeps or streams. Blackberries are typically much more tolerant of drought, flooding and high temperatures, while red raspberries are more tolerant of cold winters.

3.1.5.1 General Blackberry and Raspberry Habit

Plants produce biennial canes from a perennial crown/root system with vegetative canes, called "primocanes", growing for the first year. After a dormant period they flower and are called "floricanes" that produce fruit before dying. Simultaneous with the floricanes fruiting, the plants are producing primocanes for the following year's crop. In any given year, half the canes on a plant are vegetative primocanes and the other half fruiting floricanes. While the plants are perennial, the canes are typically biennial and the floricanes die after fruiting. Primocane-fruiting red raspberry, black raspberry, and blackberry cultivars have been developed that are able to flower and fruit on the current season's growth. All cultivated red raspberries and some blackberries produce root initials on spreading roots to form new canes so the plants can spread quite some distance vegetatively. All black raspberries and many cultivated blackberries are crown forming and do not spread underground; however, they both can spread vegetatively by tip layering of primocanes. Blackberries are generally larger and more vigorous than raspberries, and the cultivated types have prostrate (trailing) to very upright (erect) growth habits with canes up to 5 m tall (Clark et al. 2007). The primary cultivated red raspberry species is very erect while the commercial blackberries have a more varied growth habit.

3.1.5.2 Blackberry Growth Habit

While blackberry species range from completely procumbent to very upright, commercial blackberries are classified into three categories based on cane type: trailing, semi-erect, and erect (Strik 1992). Trailing types are crown-forming and the primocanes trail on the ground surface until they are bundled, lifted and tied to a trellis. "Marion", "Thornless Evergreen", and "Black Diamond" are examples of this type of plant. Those with a semi-erect habit are also crown-forming and require a trellis with the mature canes growing erect for about 1 m before arching over. As mentioned previously, important semi-erect cultivars are "Chester Thornless", "Loch Ness", and "Triple Crown". The erect blackberries grow upright, but less vigorously than the semi-erect types, and instead of being crown forming, they sucker beneath the soil line and will form a continuous row of canes in a managed field. Erect cultivars include "Navaho", "Arapaho", "Ouachita", and "Natchez". Primocanes of erect and semi-erect blackberry cultivars and black raspberry cultivars are tipped in the summer to encourage branching and to increase production, whereas trailing cultivars and red raspberries are usually not tipped.

3.1.5.3 Raspberry—Red, Black, and Purple—Growth Habit

Red raspberries sucker vegetatively beneath the soil producing new canes from root initials. The primocanes of many raspberry cultivars remain vegetative in the first year and are only harvested in the second year as floricanes. Some of the more popular cultivars of this type are "Tulameen", "Glen Ample", "Meeker" and "Willamette". There are also primocane fruiting red raspberry cultivars that produce fruit in the fall at the top of the current season's canes and then again in the second year at the base of the cane. Some of the more popular cultivars of this type include "Heritage", "Caroline", "Josephine", "Amity" and "Autumn Bliss". While it is easiest to cut the canes of these cultivars off at ground level each winter after picking the late-summer primocane crop, the canes are sometimes left to over-winter and produce a very early spring crop. Because these primocane fruiting types can be double cropped, they are sometimes called "everbearing raspberries".

With the exception of "Explorer", all black raspberry cultivars are floricane fruiting. The primocanes that emerge from the crown are tipped in commercial plantings to about 1 m tall to encourage branching. The following year these canes become floricanes and produce the crop. A few of the more widely planted cultivars of black raspberry are "Munger", "Jewel", and "Blackhawk". Purple raspberries, a cross between red and black raspberries, tend to have a great deal of hybrid vigor and are also crown forming plants with large, soft fruit. They are generally considered to have only fair quality freshness, but truly shine when they are processed. "Brandywine" and "Royalty" are the two most commonly sold cultivars.

3.1.5.4 Ploidy and Genome Size of the Brambles

The ploidy level in the genus *Rubus* ($x = 7$) can vary across different species form $2n = 14$ to $2n = 98$. Red raspberry is generally considered as being diploid with the exception of a few naturally occurring triploids and tetraploids (Finn 2008). Blackberry can vary from $2n = 14$ to $2n = 84$. However, the ploidy levels for the majority of genotypes that are used in commercial breeding programs are between $2n = 14$ to $2n = 28$ (Clark et al. 2007). The aforementioned hybrid "Loganberry" is a cross between a 2x raspberry and likely an 8x blackberry (*R. ursinus*) to give a ploidy of $2n = 42$. "Boysenberry" and "Youngberry" are septaploids ($2n = 49$) (Clark et al. 2007). The genome size has been estimated in red raspberry at 245 Mbp (Graham et al. 2004).

3.2 Diversity Analysis

3.2.1 Genotype-based Diversity Analysis

Many studies have utilized molecular markers to quantify genetic diversity in wild raspberry populations as well as among breeding programs and cultivars. Molecular markers have also been used in classifying diversity and elucidating the relationships among the many hundreds of *Rubus* species (Weber 2003).

Various DNA-fingerprinting techniques have been used to investigate diversity and population dynamics in raspberry. Minisatellite-based fingerprints were developed in black raspberry that identified 15 different genotypes from a sample of 20 individuals (Nybom et al. 1990). Restricted DNA was probed from 22 red and purple raspberry genotypes with chloroplast DNA probes and determined that maternal ancestry in *R. occidentalis*, *R. parvifolius* and *R. strigosus* (syn. *R. idaeus strigosus* Michx.) could be differentiated from *R. idaeus* (syn. *R. idaeus vulgatus* Arrhen.), from which arose all of the commercial red raspberry cultivars that they studied (Moore 1993).

RAPD markers have been used widely in taxonomic studies (Graham and McNicol 1995; Pamfil et al. 1997; Trople and Moore 1999; Pamfil et al. 2000) and germplasm diversity assessments (Graham et al. 1994, 1997, 2003; Weber 2003; Patamsytė et al. 2004; Badjakov et al. 2006) in raspberry species. For example, Graham et al. (1994) used the distribution of markers from nine RAPD primers to produce a similarity index and hierarchical tree for 10 red raspberry cultivars, and Graham and McNicol (1995) elucidated relationships among 13 *Rubus* species using RAPDs.

In another use of RAPD markers, Graham et al. (1997) examined relationships and spatial diversity among wild populations of *R. idaeus* in Scotland at four sites in comparison to the commercial cultivar "Glen Clova"; where none of the wild populations were closely related to the commercial cultivar. Later, Graham et al. (2003) assayed a wider range of wild *R. idaeus* from 12 sites across a greater area of the United Kingdom and compared the accessions to "Glen Moy". Again, little gene flow was observed between wild populations and commercial cultivars.

AFLPs have been used to investigate diversity in multiple *Rubus* species. Amsellem et al. (2000) investigated the weedy raspberry species *R. alceifolius* in its native range and in areas where it has invaded as a noxious weed (Amsellem et al. 2001c). Considerably more genetic diversity was detected in its native range with diversity in non-native ranges dependent on distance from the origin. Multiple introduction sites could be identified in some cases. Amsellem et al. (2001b) also used AFLPs to show that reproduction within the native range of *R. alceifolius* is through sexual means while apomixis was common only in non-native locations (Amsellem et al. 2001b). Marking (2006) also found little difference in genetic diversity between natural populations and cultivated populations using chloroplast sequence and SSRs (Marking 2006).

Lindqvist-Kreuze et al. (2003) also used AFLP markers to characterize diversity in six populations of wild arctic raspberry (*R. arcticus* L.) and 10 cultivars in Finland. AFLPs were highly effective in distinguishing 78 genotypes from 122 samples. Genetic variation was found to be high within populations, indicating a high degree of sexual reproduction, but interpopulation gene flow was low as measured by overall diversity among locations. Diversity within the cultivars was high enough that subspecies could be differentiated (Lindqvist-Kreuze et al. 2003).

3.2.2 Relationship with other Cultivated Species and Wild Relatives

Trople and Moore (1999) calculated genetic similarities among 43 Rubus species and raspberry genotypes based on marker profiles from six RAPD primers. The similarity indices were relatively low between the species (0.15 to 0.52) with much higher indices for multiple accessions within species (0.62 to 0.82). In another study, 40 species of Rubus were analyzed (including many raspberry types) using RAPD markers and found that molecular classification of species agreed with the traditional classification of Rubus in most cases, except for three species in the subgenus Malachobatus that clustered with the raspberry types in subgenus Idaeobatus (Pamfil et al. 2000). However, their RAPD based taxonomy could not explain differential success of interspecific hybridization within each subgenus.

Using RAPD markers, Badjakov et al. (2006) analyzed 28 raspberry genotypes from the Bulgarian germplasm collection including 18 Bulgarian cultivars and breeding lines, eight accessions from outside Bulgaria and two wild species accessions, *R. occidentalis* and *R. adiene*. They created a genetic similarity tree with two clusters that corresponded to two pedigree groups among the Bulgarian genotypes. They also analyzed the 28 accessions with four SSR loci, demonstrating high levels of diversity within the collection (Badjakov et al. 2006).

3.2.3 *Extent of Genetic Diversity and Relationship with Geographical Distribution*

Research with natural populations has shown genetic diversity estimates in raspberry at levels near 50% among and within population (e.g., Lindqvist-Kreuze, et al. 2003 for arctic raspberry). With RAPD markers, Graham et al. (1997) examined spatial diversity in wild accessions of *R. idaeus* from four sites in Scotland. Most of the variability in markers was observed between the collection sites. Within sites, increasing diversity coincided with greater spatial separation. None of the wild populations were closely related to the commercial cultivar. Later, Graham et al. (2003) assayed a wider range of wild *R. idaeus* from 12 sites across a greater area of the UK and compared the accessions to "Glen Moy". Again, greater genetic similarity was found within each population collected, which indicates a hindrance of gene movement across geographic locations.

At the species level, Alice and Campbell (1999) found that nuclear sequences were generally consistent with biogeography and ploidy levels and less so with morphological traits. They produced a *Rubus* phylogeny of 57 species including multiple raspberry species based on ribosomal internal transcribed spacer region (ITSR) sequence variation. The *Rubus* subgenus *Idaeobatus* of the Pacific region was studied in comparison with species from other subgenera to evaluate biogeographic and phylogenetic affinities of *R. macraei* A. Gray, using chromosome analysis and chloroplast gene *ndhF* sequence (Morden et al. 2003). Their results showed that *R. macraei* is more similar to the blackberry species of the subgenus *Rubus*. Moreover, they discovered that *R. macraei* and *R. hawaiensis* A. Gray are derived from separate colonizations from North America and that similarities between them are due to convergent evolution in the Hawaiian environment.

Twenty wild *R. idaeus* accessions from a Lithuanian germplasm collection were examined for genetic diversity using 285 RAPD loci produced from 36 primers (Patamsytė et al. 2004). Genetic distances among the genotypes did not correlate to geographic distances between collection sites; however, soil acidity was significantly correlated to observed polymorphisms in the RAPD markers, indicating an environmental effect on diversity within populations.

A study on 63 natural populations of *R. strigosus* across North America (Marking 2006) using chloroplast sequence and interlocus-simple sequence repeats (ISSR) found 79.5% of the total variation to be within populations rather than between the populations studied. The total genetic diversity (π, which is defined as the average number of nucleotide differences per site between any two DNA sequences chosen randomly from the sample population) was found to be $\pi = 0.009$ for the chloroplast sequence data, with the genetic variation among populations at $\pi = 0.001$. Expected heterozygosity levels from the nuclear ISSR marker data ranged from 0.84–1.00. Population specific Fst values (a statistic that compares the genetic variability within and between populations) ranged from 0.183–0.271, with an average Fst value of 0.208 among all populations. Marking found no evidence of correlations among geographic location and the level of genetic diversity nor any evidence of increased diversity in naturally existing populations compared to the cultivated populations of North American red raspberry. Weber (2003) analyzed genetic diversity in cultivars of black and red raspberry using RAPD markers and found that black raspberry genotypes showed on average 81% genetic similarity; this compared to 70% similarity measured among red raspberry cultivars in Europe (Graham et al. 1994). Of the 16 genotypes investigated, five cultivars accounted for 58% of the observed variability in black raspberry, and none of the black raspberry cultivars were more than two generations from at least one wild ancestor.

DNA probes from two variable number tandem repeat (VNTR) loci were utilized to examine diversity in wild populations of *R. moluccanus* L. in the Philippines (Busemeyer et al. 1997). The results were much the same to that of Graham et al. (1997, 2003) in *R. idaeus*, more similarity was found to be present within populations at each location than between locations. Additionally, apomictic reproduction was ruled out in these populations because no identical VNTR patterns were identified.

3.3 Classical Genetics and Conventional Breeding

3.3.1 Classical Mapping Efforts

Early work on linkage analysis of morphological traits by Crane and Lawrence (1931) and Lewis (1939) documented aberrant segregation ratios among populations segregating for fruit color (T) and pale green leaves (g or ch$_1$) in red raspberry (Crane and Lawrence 1931; Lewis 1939). Further research showed genetic linkage among five genes (waxy bloom *b*, apricot or yellow fruit *t*, pale green leaf *g*, red hypocotyl *x* and pollen tube inhibitor *w*), producing the first genetic linkage group for *Rubus* (Lewis 1939, 1940). Sepaloid *sx*$_3$ was later added to the linkage group between *b* and *t* (Keep 1964). Crane and Lawrence (1931) and Lewis (1939, 1940) also postulated on a linkage between a semi-lethal allele with the unlinked *h* gene. Jennings (1967) added further evidence to this linkage, proposing the symbols *wt* for the locus linked to the fruit color *t* locus and *wh* linked to the hairy locus (*h*) in place of *w* that Lewis (1939) used (Jennings 1967).

Subsequent work in red raspberry has further elucidated the inheritance of hairiness and fruit color as well as numerous other traits. Associations between the *H* allele for cane hairiness and resistance to spur blight (*Didymella applanata* [Niessl] Sacc.), cane botrytis (*Botrytis cinerea* Pers.:Fr.)

and cane blight have been recognized (Jennings 1988; Keep 1989). This same gene has also been associated with susceptibility to anthracnose cane spot (*Elsinoe veneta* [Burkholder] Jenk.), powdery mildew (*Sphaerotheca macularis* [Wallr.:Fr.] Lind.) and western yellow rust (*Phragmidium rubi-idaei* [DC.] P. Karst.) (Jennings and McGregor 1988; Jennings and Brydon 1989). These associations are likely pleiotropic effects rather than genetic linkages. Similarly, the recessive gene *s* for spine-free canes and the dominant *B* for waxy bloom on canes can reduce spur blight incidence (Jennings 1982b, 1988). Again, it is unlikely these genes are genetically linked but rather complement each other physiologically to reduce disease incidence. However, no other linkage groups based solely on morphological traits have been proposed. Daubeny (1996) lists 72 individual loci or alleles that have been identified, many of which are part of an allelic series for aphid resistance (Daubeny 1996). Corresponding work in blackberry and other *Rubus* species has been largely absent, probably due to the complex genetics of blackberry and the small number of breeding programs devoted to them.

3.3.2 *Limitations of Classical Endeavors and Utility of Molecular Mapping*

Prior to the advent of molecular markers, inheritance and genetic mapping studies were limited to simple morphological traits (Ourecky 1975; Jennings 1988). These studies generally utilized phenotypes that are deleterious in the recessive form so that they are undesirable to maintain in breeding programs. The advent of biotechnology has resulted in a fundamental shift in the development of genetic linkage maps and their use in cultivar development. Classical breeding, which selects parents and their desirable offspring based on an observable phenotype, is being integrated with techniques that can identify and manage genetic variability at the molecular level (protein or DNA). The ability to detect genome wide variability has led to the characterization of genetic variation within not only coding regions (i.e., genes and their morphological manifestations) but in non-coding regions as well, which make up large portions of plant genomes. These developments have enabled the construction of genetic linkage maps of red raspberry containing numerous genetic markers that are phenotypically neutral that have been used to identify genomic regions associated with phenotype (see below).

3.3.3 *Breeding Goals and Objectives*

While all breeding programs share some common goals, they also have specific objectives depending on the type of raspberry or blackberry grown in their region, the use and market for the fruit, and genetic variability available (Clark and Finn 2006; Table 3-1). In general terms, yield, fruit quality, abiotic and biotic stress tolerance, and harvest ease are all important traits that have been well addressed by breeders.

Fruit quality has been identified as a primary focus of all programs as it is the main area that can increase blackberry and raspberry consumption (C Finn and J Clark, pers.comm.). Advances in fruit quality from the early wild selections and first improved cultivars have been substantial. In blackberry, this progress has moved it from being viewed as a fruit harvested from the wild that was often not found on grocers' shelves to one that is now found on shelves year round and from all parts of the world. In the fresh market, cultivars with improved flavor and sweetness are foremost in need of enhancement. Progress in this area can be made along with manipulation of flavor components, acidity, astringency, and postharvest handling. There is more than adequate genetic variation available to advance blackberries substantially further, especially in quality attributes. It is hoped that this increase in quality cultivars may displace current cultivars which may deter repeat sales by consumers. Similarly with raspberries, the first goal in breeding for fruit quality of fresh products was to develop genotypes with much firmer fruit that would allow the delicate raspberry to be harvested and shipped around the world. Unfortunately, this has led to the development of very firm berries that are picked immature at the expense of taste. More recently, flavor has been reincorporated into commercial red raspberries without a substantial compromise in fruit firmness or shipability. Cultivars for processing generally have intense flavor and high soluble solid levels. The challenge for breeders is to maintain or improve these characters while thornlessness, machine harvestability, increased yield, and increased fruit firmness are selected for in breeding populations.

Breeding for broader adaptability or for types that will do well in new regions has greatly expanded the industry. In raspberry, California has become the predominant producer due to the development of superior primocane fruiting types that work on an 18 month production cycle. Blackberry production has dramatically expanded in areas with very low chilling, particularly Mexico. While production techniques including defoliation and the use of growth regulators has been a major part of this development in Central America, México, the use of the Brazilian cultivar Tupy, has been equally important. Further, the recent introduction of primocane fruiting in blackberry offers a method to eliminate chilling concerns completely, since these canes do not go through a dormant period prior to initiating flowers.

Adaptation to heat has been a major problem for raspberries and blackberries. While blackberries are generally better adapted to heat than raspberries, high temperatures during fruit ripening can be equally devastating to each crop. While there is genetic variability for this trait, it has not been well characterized.

Overall, variability for most traits such as thornlessness, architecture, disease and insect resistance, adaptation, productivity, fruit quality and size, and other traits is sufficient enough that progress can be made if a breeding program focuses on that trait and can assemble the appropriate germplasm. The exception to this is in black raspberry where there is very little variability (Dossett et al. 2008; Weber et al. 2008).

3.3.4 Achievements of Conventional Breeding

Blackberries and raspberries have a relatively short history as cultivated crops that have been enhanced through plant breeding. Further, an intensive focus on plant improvement has taken place for less than a century and modern cultivars are very few generations removed from their wild progenitor species. The improvements that have allowed these plants to be commercially cultivated crops are well documented, including increased yield, improved harvest efficiency, abiotic and biotic stress tolerance, increased fruit quality for fresh and processed markets, altered plant architecture, etc. There are excellent reviews on blackberry and raspberry breeding, genetics and germplasm (Darrow 1937; Waldo 1950, 1968; Oydvin 1970; Sherman and Sharpe 1971; Moore 1984; Jennings 1988; Hall 1990; Jennings et al. 1991; Daubeny 1996; Clark and Finn 2008; (CE Finn and JR Clark, pers. comm.); (C Kempler, pers. comm.).

3.4 Linkage Mapping and Molecular Breeding

3.4.1 A Brief History of Marker Development

As mentioned previously, the first linkage map for *Rubus* was presented by Lewis in 1939. He postulated linkage between the genes controlling five traits, namely B (bloom on primocanes), T (presence of red pigments in fruit and spines), G (normal vs. pale green leaf), X (green vs. red hypocotyls) and W (responsible for selective fertilization), based on aberrant ratios in a series of more than 60 different progenies of red raspberry using his own data and that of Crane and Lawrence (1931) (Crane and Lawrence 1931; Lewis 1939, 1940). Subsequently, linkage was reported between gene H, controlling cane pubescence and resistance to cane botrytis and spur blight as well as increased susceptibility to yellow rust, anthracnose cane spot and powdery mildew in red raspberry (Knight and Keep 1958; Keep and Knight 1968; Jennings 1982a, b). Similarly, Jennings and Keep reported linkages between dwarf habit phenotype (*dw*) and genes H and T first described by Crane and Lawrence in 1931 (Crane and Lawrence 1931; Jennings 1967;

Keep 1968). We would have to wait until the advent of molecular markers before linkage maps spanning the genome can be developed and some of these hypotheses could be proven or refuted.

3.4.2 Molecular Marker Evolution and Development in Rubus

A number of molecular markers have been employed in the genetic analysis of raspberry, including RAPDs, AFLPs and microsatellites or simple sequence repeats (SSRs). RAPDs have been used for linkage map construction (Pattison et al. 2007) to determine the relationships within *Rubus* species (Graham and McNicol 1995; Stafne et al. 2005) and to assess the genetic diversity of *Rubus* species such as black raspberry (Weber 2003) where a lack of genetic variation is a limiting factor in the production of new commercial cultivars and wild *R. idaeus* (Graham et al. 1997). RAPDs have also been used to identify raspberry cultivars (Parent and Page 1992). Likewise, AFLPs have been employed for linkage map construction (Graham et al. 2004b; Graham et al. 2006; Pattison et al. 2007; Sargent et al. 2007; Woodhead et al. 2008) as well as to assess the genetic diversity of wild and cultivated *Rubus* species (Amsellem et al. 2000; Lindqvist-Kreuze et al. 2003; Marulanda et al. 2007).

However, in recent years, due to their well documented advantages for diversity studies and linkage map construction, SSRs have become the marker of choice for genetic analysis in many crop species, including raspberry. Microsatellites have been developed from a number of different raspberry species, including *R. alceifolius* (Amsellem et al. 2001b), and from *R. idaeus* (Graham et al. 2002, 2004b, 2006; Lopes et al. 2006). Recently, EST-SSRs have been developed in raspberry and blackberry that have the advantage of providing functional information at that locus (Lewers et al. 2008; Woodhead et al. 2008).

In addition, SSR markers derived from genera closely related to *Rubus*, such as *Fragaria*, have been applied to *Rubus* and have been shown to be transferable and polymorphic, thus suitable for genetic analysis (Lewers and Hokanson 2005a; Stafne et al. 2005).

3.4.3 Mapping Populations Used

The most saturated *Rubus* map currently available was constructed from a population derived from a cross between two subspecies of red raspberry, "Glen Moy", predominantly *R. idaeus* and "Latham", predominantly *R. strigosus* (although some *R. idaeus idaeus* is existent in "Latham's" background) (GM×L), by Graham et al. (2004). The population segregates for a number of phenotypic characters including root-rot resistance. Three maps so far have been published on this progeny set. Graham et al. (2004)

employed SSR and AFLP markers to define a linkage map comprising 273 markers covering 789 cM over nine linkage groups, and quantitative trait loci (QTLs) were identified for spine density and two measures of plant vigor, the density and spread of root suckers. Graham et al. (2006) later added 20 SSR markers to the GM×L map along with analyzing data on the *H* gene for cane hairiness and resistance to multiple fungal pathogens. This map contained 349 markers across a total length of 669 cM and eight linkage groups. The *H* gene was mapped to linkage group 2 and associated closely with resistance to cane botrytis and spur blight. Resistance to cane spot was mapped to a different region of linkage group 2 and linkage group 4, and yellow rust resistance was mapped to linkage group 5. The latest GM×L map (Woodhead et al. 2008) introduces a set of 25 SSRs derived from EST sequences from raspberry root and bud tissues. This map covers a total of 884 cM in seven linkage groups and highlights QTL position for fruit quality attributes.

A number of other linkage maps have been reported for *Rubus* and contain map positions for resistance to pests and diseases of key economic importance in *Rubus*. Pattison et al (2007) raised F_1, F_2, BC1, BC2 and S_1 populations from a cross between the "Latham" and "Titan" and used these progenies to study the inheritance of resistance to Phytophthora root rot (PRR) in *R. idaeus*. A dominant two gene model was shown to be the best fit for the segregation for resistance observed, and a molecular linkage map was constructed of both parents from their B2 population, incorporating 138 AFLP, 68 RAPD, 20 resistance gene analogue markers. The linkage map spanned seven linkage groups covering 440 cM in "Latham" and 370 cM in "Titan". QTL analysis of the B2 population highlighted two genomic regions on each map that were associated with PRR resistance. Bulked segregant analysis (BSA) corroborated this conclusion by identifying markers from these regions associated with bulked samples of resistant and susceptible genotypes. Generational means analysis suggests two major genes controlling resistance possibly corresponding to the two regions on each parental linkage map associated with resistance. Unfortunately, this map was not aligned to the LxGM reference map.

Sargent et al. (2007) constructed a linkage map of the progeny "Malling Jewel" × "Malling Orion" (MJ×MO) that segregated for the gene A_1 which confers resistance to the aphid *Amphorophora idaei* and for *dw*, a gene for dwarfing habit in *R. idaeus*. The map was produced from an F_1 progeny of the cross MJ×MO and contained 95 AFLP and 22 SSR markers in seven linkage groups and covered a total of 505 cM. The two phenotypic traits were mapped to linkage groups 3 and 6 (Graham et al. 2006) and were closely flanked by SSR and RAPD markers.

3.5 Mapping of Simply Inherited Traits

3.5.1 Aphid Resistance

The large raspberry aphids, *Amphorophora idaei* Börner in Europe and *A. agathonica* Hottes in North America, transmit several viruses including *Raspberry leaf mottle virus* (RLMV), *Raspberry leaf spot virus* (RLSV), *Black raspberry necrosis virus* (BRNV) and *Rubus yellow net virus* (RYNV); thus, breeding for vector resistance is a key goal of many breeding programs. The genetic control of *A. idaei* resistance derived from different sources was elucidated, attributing it to series of single dominant genes (A_1- A_{10}, A_{L518}, A_{K4a}) (Keep and Knight 1967; Keep et al. 1970). However, resistance is strongly dependent upon the *A. idaei* biotype with some genes providing resistance to certain biotypes but not others (Sargent et al. 2007). The widespread growing of cultivars carrying A_1 resistance imposed a strong selection pressure on aphid populations causing biotypes that could overcome this resistance to become predominant in the UK during the 1980s and 1990s (Birch et al. 1997). Resistance to the most common biotypes can be conferred by one of three genes: A_{10} from *R. occidentalis*, A_{L518} from *R. strigosus* or (A_{K4a}) from a German clone of *R. idaeus*. However, an A_{10} resistance-breaking biotype has been reported (Birch et al. 1997) making it ever more important to pyramid several aphid resistance genes in breeding lines to provide a robust and durable vector resistance. Suitable molecular markers are essential to determine how many resistance genes are carried by resistant selections as the equivalent effects of many of the reported resistance genes make them phenotypically indistinguishable and possibly synonymous. Thus the development of maps of progenies carrying resistance genes and the identification of molecular markers linked to these genes will provide a key tool in differentiating reported genes, identifying their presence in modern hybrid material, and managing strategies for pyramiding. As stated previously, A_1 has been mapped to LG3 (Sargent et al. 2007), and linked markers to the other genes are needed to enable efficient gene pyramiding.

3.5.2 Dwarf Habit and Pubescent Canes

Jennings (1967) reported and depicted a dwarf phenotype that he attributed to the recessive gene *dw* and speculated that *dw* was linked with *H* and *T*, the genes for pubescent stems and for presence or absence of anthocyanin. Keep (1968) attributed this phenotype to the effect of two genes, *d1* and *d2*, and accepted Jennings's hypothesis of linkage between this character and the *H* and *T* genes.

Keep (1968) proposed that some dwarf phenotypes could have some interest for self-supporting breeding lines; however, because of other agronomic disadvantages such as reduced fertility and longevity, this has not been pursued. Moreover, this recessive gene is still present in modern breeding lines, and it becomes apparent in breeding progenies every so often. Therefore, it could be useful to include a linked marker to parental screenings for marker-assisted breeding (MAB) thus avoiding unfavorable crosses. The importance of gene *H* (cane pubescence) is related to its linkage with cane disease resistance traits. Mapping has confirmed its linkage to QTLs for resistance to two important cane diseases, namely spur blight and cane botrytis (Graham et al. 2006) as had been previously reported (Knight and Keep 1958; Keep 1968; Jennings 1982a, b). However, despite breeders' observations suggesting otherwise, hairy canes did not show association with susceptibility to cane spot or yellow rust in the LxGM progeny. The hypothesis of linkage between *H* and *dw* has also been revised as *H* maps to LG2 (Graham et al. 2006) where as *dw* was mapped to LG6 (Sargent et al. 2007).

3.6 Molecular Mapping of Complex Traits

3.6.1 Root Rot Resistance

Raspberry root rot, caused by *Phytophthora* species, is a devastating soil-borne disease that has caused serious problems in Europe, North America and Australia for many years, and screening for resistance has been ongoing from the late 1970s (Barritt et al. 1979; Bristow et al. 1988; Duncan and Kennedy 1992; Knight and Fernández-Fernández 2008). Several breeding programs have resistance to root rot as a major breeding target, thus robust and transferable molecular markers to allow reliable identification of genotypes with high levels of resistance to this disease would be very beneficial to breeders. Efforts on both sides of the Atlantic are being devoted to this aim. Pattison et al. (2007) presented an elegant two gene hypothesis for the genetic control of the trait that was mapped to two different linkage groups and ongoing work by (J Graham, pers. comm.) identified SSR markers linked the QTLs for resistance identified in the GM×L population as well as assessing their transferability to other genetic backgrounds. In view of these parallel efforts, it would be beneficial to align both maps.

3.6.2 Resistance to Cane Diseases

Cane botrytis and spur blight infect mature or senescent leaves on raspberry primocanes. The infection spreads to the buds where the disease lesions develop. The following spring, buds from infected nodes will break later or not at all constituting a major cause for yield loss, particularly in the case of cane botrytis. A common resistance mechanism to both diseases was proposed (Williamson and Jennings 1986), and Graham et al. (2006) identified two major QTLs associated with both resistances in LG2 and LG3 of the LxGM progeny suggesting multiple gene control of the trait. Interestingly, the QTL in LG2 collocated with gene *H* previously discussed.

Cane spot or anthracnose (*Elsinoe veneta* [Burkh.] Jenkins,) can develop in most raspberry tissues, but it is most recognizable in the second year canes, where it produces deep lesions that can lead to vascular damage and therefore reduce yields. The resistance to this pathogen has been associated to the presence of hairy cane (*H*) in European red raspberry but not so in North American cultivars (Graham et al. 2006). Genetic control of the trait has not yet been firmly established, although Graham et al. (2006) identified two QTLs in LG2 and LG4 of the LxGM progeny associated with response to the disease.

Yellow rust (*Phragmidium rubi-idaei* [DC.] P. Karst.,) has increased its prevalence in recent years as the cultivation under tunnels of susceptible cultivars (e.g., "Glen Ample" and "Tulameen") has become common. A major resistance gene (*Yr*) from "Latham" was identified by Anthony et al. (1986), and Graham et al. (2006) postulated this gene to be located in LG3 of the GM×L progeny. The inheritance of complete and incomplete resistance to rust in a half diallel cross including "Boyne", which derives complete resistance from "Latham", was studied (Anthony et al. 1986). They found that crosses of "Boyne" to susceptible cultivars all segregated for complete resistance and proposed that "Boyne" was heterozygous for a single resistance gene, designated *Yr*, which was derived from "Latham". As the GM×L population also segregates for resistance to rust, Graham et al. (2006) proposed that "Latham" is also heterozygous for *Yr* and that this lies on linkage group 3 close to E41M31-147. Anthony et al. (1986) also found variation in the degree of susceptibility among offspring of "Boyne" without complete resistance and concluded "Boyne" to also be a source of incomplete resistance.

In the GM×L cross, there is some evidence, although not highly significant, for a gene on linkage group (LG) 5 (also from "Latham") affecting the susceptibility of the offspring that do not carry the "resistant" allele on linkage group 3. This area on LG5 is also implicated in spur blight/botrytis resistance. There was no evidence, however, of gene *H* being related to incomplete resistance in this cross. None of the offspring in this cross were as susceptible to rust as the "Glen Moy" parent. One explanation of this is that there is another resistance gene for which "Latham" is homozygous *RR* and "Glen Moy" is *rr*. In this case, all offspring would be *Rr* and more resistant than "Glen Moy".

3.6.3 Viral Resistance

Attempts to develop markers for other viral resistance genes have been carried out for *Raspberry leaf spot virus* (RLSV) and *Raspberry vein chlorosis virus* (RVCV). Field screening was carried out to measure symptom production of RLSV and RVCV in two different environments. These traits were analyzed for significant linkages to mapped markers and resistance loci were found on linkage groups 2 and 7 (Rusu et al. 2006).

An alternative approach has been suggested to detect resistance to *Raspberry busy dwarf virus* (RBDV) (MacFarlane and McGavin 2009). Essentially EST clones of the virus were inoculated into red raspberry genotypes and infection was recorded. These studies found that the virus genome was greatly activated when certain regions of the RBDV coat protein was included in the inoculums.

3.6.4 Fruit Quality Traits

Progress towards mapping fruit quality traits related to eating quality and their health benefits is underway. Data have been collected on anthocyanin content across seasons and in different environments (Kassim et al. 2009). Here high performance liquid chromatography (HPLC) was used to quantify eight major anthocyanins, cyanidins, and pelargonidin glycosides: -3-sophoroside, -3-glucoside, -3-rutinoside and -3-glucosylrutinoside, in progeny from the GM×L cross. The locus associated with abundance of eight antioxidants mapped to the same chromosome region on LG1 of the map of Graham et al. (2006) across both years and from fruits grown in the field and under protected cultivation. Seven loci associated with antioxidants were mapped to a region on LG4 across years and for both field and protected sites. Candidate genes that may underlie anthocyanin production including bHLH (Espley et al. 2007), NAM/CUC2 (Ooka et al. 2003) like protein and bZIP transcription factor (Holm et al. 2002; Mallappa et al. 2006) were also identified. Volatiles have also been quantified across sites and seasons, and mapping is underway (A Kassim pers. comm.). Color has been measured visually and instrumentally and mapped to three linkage groups (J McCallum pers. comm.). Sugars and acids as well as sensory data have been collected from the GM×L population, and mapping is underway (D Zait pers. comm.). If QTLs are identified, these data and resulting markers may be integrated into existing breeding programs and could allow for early selection of important high priority fruit quality traits.

Work towards the genetic mapping of health-related compounds has been initiated in *Rubus*. With the emergence of metabolomics, the simultaneous analysis of multiple metabolites at specific time points is now feasible. In *Rubus,* a metabolomic approach has been used to identify bioactive compounds in a segregating mapping population planted under two different environments (Stewart et al. 2007). As a greater understanding of the relative importance and bioavailability of the different antioxidant compounds is achieved, it may become possible to develop and identify those raspberry genotypes with enhanced health-promoting properties from breeding programs (Beekwilder et al. 2005).

3.6.5 *Future Prospects and Work in Progress*

Ongoing work at the Scottish Crop Research Institute (UK) aims to further saturate their very well characterized, replicated progeny (GM×L) to map fruit quality characters' attributes. A candidate gene approach (Woodhead et al. 2008) is being combined with extensive phenotypic measurements, including metabolomic profiling (Stewart et al. 2007), to identify quality traits and anthocyanin transcription factors (Kassim et al. 2009). Work is also focusing on identification of gene H through BAC library screening (J Woodhead pers. comm.) and chromosome walking, and this sequence characterization is also being applied to the root rot resistance QTL (J Graham unreported data).

At East Malling Research (UK), progress is been made to map further aphid resistance genes. The "Autumn Bliss" × "Malling Jewel" population (ABxMJ), currently being tested with SSRs and RAPDs, segregates for the A_{10} gene of resistance to the large raspberry aphid *A. idaei* (F Fernández-Fernández et al. unpublished data). These progeny also segregate for primocane fruiting; a trait of great economical importance normally attributed to a combination of major and minor genes. Further progenies are being developed to map A_{L518} and A_{K4a} (F Fernández-Fernández et al. unpublished).

Work carried out at North Carolina State University (US) is using the F_1 cross (*R. parvifolius* × "Tulameen") × "Qualicum" to elucidate the genetics of *Rubus* response to environmental conditions. *Rubus parvifolius* is of interest to breeders in a time of climate change as a germplasm donor of chilling requirements (allowing germplasm to withstand fluctuations in winter temperature) and, even more interestingly, as a donor of heat tolerance. Both traits and several others of horticultural interest segregate in the (*R. parvifolius* × "Tulameen") × "Qualicum" progeny that is currently being analyzed with AFLP and SSR markers (Molina-Bravo unpublished data).

All efforts to date have concentrated in red raspberry, however, work is being carried out at Cornell University on a black raspberry mapping progeny (C Weber unpublished data). Extensive work in black raspberries is also taking place with the USDA-ARS in Oregon, with effort in mapping genes for aphid resistance and regional adaptaption (M Dossett, pers. comm.). We can also expect blackberry, at least in its tetraploid form, to become a subject of mapping work in the near future as more resources become available for these species, such as the recent EST library for SSR development published by Lewers et al. (2008).

3.7 Molecular Breeding

3.7.1 Germplasm Characterization

The first attempts to "fingerprint" *Rubus* germplasm using paper chromatography (Haskell and Garrie 1966) were followed by isoenzyme analyses (Cousineau and Donnelly 1989) and more recently by DNA based markers such as minisatellite DNA (Nybom et al. 1990) and RAPDs (Parent and Page 1992; Weber 2003). With their increasing availability for the *Rubus* genus, SSRs are becoming the most commonly used fingerprinting tool, and standardized sets of SSRs are needed for use in each crop. A set of *Rubus* SSRs has been identified for use in an EU project, GENBERRY (Denoyes-Rothan et al. 2008). Similar sets have been proposed for other *Rosaceous* crops (Govan et al. 2008) and when combined with a set of control cultivars, like in the case of *Pyrus* (Evans et al. 2007), it allows researchers to readily compare data. The use of markers in breeding to screen germplasm to increase diversity is important in raspberry as five parent cultivars dominate their ancestry. These cultivars are "Lloyd George" and "Pynes Royal" entirely derived from the European sub-species, and "Preussen", "Cuthbert" and "Newburgh" derived from European and North American subspecies. Domestication has resulted in a reduction of morphological and genetic diversity in red raspberry (Haskell and Garrie 1966; Jennings 1988), with modern cultivars being genetically similar (Dale et al. 1993 ; Graham and McNicol 1995). Extensive genetic diversity has been found in wild raspberry germplasm, offering scope for expanding the genetic base of cultivated raspberries (Graham et al. 1997; Marshall et al. 2001) and sourcing locally adapted material may become increasingly important in the light of climate change and other contemporary challenges.

3.7.2 Marker-Assisted Gene Introgression and Gene Pyramiding

Gene pyramiding has been successfully achieved for multiple resistances to the aphid, *A. idaei*. Segregation data for over 500 breeding progenies over the last 25 years indicates the presence of at least three independent major aphid resistance genes in the East Malling Research breeding material (F. Fernández-Fernández unpubl data). This has been achieved as much thanks to chance as to a rigorous selection for resistant germplasm. Tightly linked markers are needed to improve efficiency and ascertain the nature of the resistance in breeding lines. This characterization will allow researchers to evaluate the response of resistance breaking aphid biotypes to germplasm carrying a combination of resistance genes and therefore plan future breeding strategies. Similarly, the identification of markers linked to root-rot resistance from sources other than "Latham" would enable a gene pyramiding strategy for this disease. In Oregon, the USDA-ARS has identified two new genes that confer aphid resistance, and hopefully therefore resistance to aphid borne viruses, in black raspberry that are likely to be useful in red raspberry breeding too (M Dossett, unpub data).

3.7.3 Limitations and Prospects for Marker-Assisted Breeding (MAB) in Rubus

At present, there are only a limited number of molecular markers available for *Rubus*, and only a small proportion of those have been shown to be linked to traits of economic importance. Therefore, current molecular tools for practical plant breeding and marker-assisted selection in *Rubus* are limited.

However, as more and more data are generated from EST libraries for *Rubus* and with the genome sequencing efforts currently underway for *Prunus*, *Malus* and *Fragaria*, there promises to be a wealth of data available for candidate genes that can be used to develop gene-specific markers for *Rubus*. The number of raspberry sequences is increasing rapidly as efforts are underway to sequence EST libraries generated from different tissues and developmental stages. At the SCRI, cDNA libraries have been generated from leaves (approximately 6,500 clones), canes (approximately 8,000 clones) and roots (approximately 7,300 clones), and further libraries are being constructed from fruit and shoots (J Graham, K Smith, M Woodhead and J McCallum, unpubl data). As well as providing sequence information on genes expressed in these tissues, these resources are being used to identify DNA markers (EST-SSRs and SNPs) for use in the genetic mapping programs.

As many traits in *Rubus* have been shown to be under the control of single or a small number of genes, there is the real possibility of using these gene sequence data to generate markers for traits such as pest and disease resistance, and thus permit the tracking and pyramiding of large numbers of useful genes into single breeding lines or cultivars.

3.7.4 Transgenic Breeding

Genetic transformation is an important component of functional gene analysis studies and holds tremendous potential for crop improvement. The first successful transformation of red raspberry expressing the β-glucuronidase (GUS) marker gene was produced from leaf disks and internodal segments using *Agrobacterium tumefaciens* (Graham et al. 1990). One of the major hurdles to successful transformation was the discovery that kanamycin inhibited organogenesis in raspberry; therefore, the selectable marker gene *nptII* was a poor marker (Graham et al. 1990; Hassan et al. 1993). Later, *A. tumefaciens* transformations of the raspberry cultivars "Meeker", "Chilliwack", and "Canby" were completed with the gene S-adenosylmethionine hydrolase (SAMase). SAMase lowers ethylene production and hence could potentially increase shelf life of fruit (Mathews et al. 1995). These transformations used hygromycin phosphotransferase (*hpt*), which conferred resistance to hygromycin and lead to a reported 49% transformation efficiency in "Meeker".

In an attempt to increase yields, the *defH9-iaaM* auxin-synthesizing or parthenocarpic gene using *nptII* as a selectable marker was inserted into "Ruby" using *A. tumefaciens* with the goal of improving the productivity of raspberry (Mezzetti et al. 2002). Greenhouse trials showed a significant increase in fruit size and yield over two harvest seasons, and subsequent field testing showed increases in flowers per plant and per inflorescence translating to a 100% increase in overall yield (Mezzetti et al. 2004).

To improve virus resistance in raspberry, "Meeker" was transformed using *A. tumefaciens* with six constructs based on the coat protein and movement protein genes of RBDV, the causal agent of crumbly fruit in raspberry (Martin and Mathews 2001). Grafting tests with infected material were performed to test for virus resistance and resulted in 53 of 141 transgenic lines remaining virus free for two rounds of grafting. Additional lines were eventually developed using these genes as well as nontranslatable RNA of RBDV (Martin et al. 2004), which resulted in five lines of 197 remaining RBDV free for five years in field testing with heavy disease pressure.

While most transformations reported in *Rubus* have made use of the binary vector *A. tumefaciens* system, alternative methods have been attempted. "Stolicznaya" was transformed by directly introducing a donor plasmid into cells of a callus cell suspension with a 20% glycerol concentration (Friedrich and Váchová 1999). The plasmid carried the isopentenyltransferase gene that increases cytokinin production, and subsequent testing of callus cells on media containing no cytokinins indicated transgenic lines. Unfortunately, no plants were developed from the lines for further testing.

To date, most successful transformations have taken place in red raspberry; however, the arctic raspberry, *R. arcticus*, was transformed with the *gus-int* gene using *A. tumefaciens* (Kokko and Kärenlampi 1998). This was the first and only reported example of successful transformation of any other *Rubus* species other than red raspberry.

While regeneration systems have been developed for blackberries (Swartz and Stover 1996; Meng et al. 2004), no transgenic blackberries have been produced to date, and none of the active breeding programs have activities in this area. Meng et al. (2004) were able to achieve the highest regeneration efficiency (70% of explants) when leaves were incubated in TDZ pretreatment medium for three weeks before culturing them on regeneration medium (Woody Plant Medium with 5 uM BA and 0.5 uM IBA) in darkness for a week and then transferred them to a 16 hour light photoperiod at 23°C for 4 weeks.

As these studies have shown, there is clear potential for improving raspberry cultivars using genetic modification techniques, especially for disease resistance and yield improvements. However, by public opposition to the use of genetically modified organism (GMO) in the human food industry is limited thus preventing the realization of this technology to its full potential. For the time being, future use of these techniques will be limited to the identification and isolation of useful genes to aid breeding/ improvement purposes. In the future, using current clonal propagation systems coupled with techniques such as "Gene Deletor Technology" (Luo et al. 2007), (which effectively uses the *Cre/Lox* system to splice out transgenes from the flowers, and subsequent fruit, of the plant), the possibility exists, that genetically modified *Rubus* may be used to combine high value traits into top performing genotypes that produce transgene-free fruit.

3.8 Genomics

3.8.1 Genomic Resources

Initial work to identify genes responsible for important economic traits has come from a number of different avenues. As we have indicated previously, there has been some success identifying genes in terms of a breeding sense; however, combining traits still remains difficult to achieve, especially in high-ploidy level cultivars (Clark et al. 2007). Additionally, there has been some success in molecular mapping studies to identify QTL and marker-trait associations that have been discussed previously. Briefly, those that have been published include QTL for spines, root sucker spread and density, gene *H* (associated with cane pubescence and resistance to cane botrytis, spur blight, rust, and cane spot), phytophthora root rot resistance, dwarfing, and aphid resistance (Graham et al. 2004b, 2006; Pattison et al. 2007; Sargent et al. 2007). In order to identify the specific genes responsible for these traits, identification through map based cloning or identification through expressed sequence tag (EST) libraries will be required. Once identified, genes will need to be analyzed using correlational, mutational and complementation studies in order to elucidate specific gene functions. Additionally, these specific genes may contain single nucleotide polymorphisms (SNPs) or other marker types that may be superior to conventional markers as they lie directly on the gene. Additionally, gaining an understanding of the molecular mechanisms responsible for these traits will enable more targeted breeding in the future.

The number of *Rubus* sequences present on the NCBI database (*http:// www.ncbi.nlm.nih.gov/*) has increased rapidly recently with the number of sequences for *Rubus* spp. increasing from 822 DNA sequences (Jan 2008) to 3,862 nucleotide sequences and 310 protein sequences (April, 2010). Of the 822 sequences identified in Jan 2008, 414 were from *R. idaeus*, most of which contained SSR sequences. Of the remaining species, most sequences were evenly spread among them and were genes that were used traditionally for taxonomy and included *ndhf*, tRNA-Leu (trnL), and ITS 5.8S ribosomal RNA (Alice and Campbell 1999). When this search is compared to the April 2010 search, of the 3,862 sequences, 602 sequences are from *R. idaeus*, and 2,678 are ESTs from *R. ulmifolius* var. *inermis* × *R. thyrsiger* with the remainder being evenly spread among the other *Rubus* spp. (See Table 3-2). This large increase of 2,678 sequences from *R. ulmifolius* var. *inermis* × *R. thyrsiger* is due to the creation and extensive sequencing of an EST library at the USDA-ARS aimed toward SSR marker development (Lewers et al. 2008).

3.8.2 Genomic Libraries

Genomic libraries can act as a bridge to allow physical mapping, positional cloning of individual genes, and a scaffold for whole genome sequencing. *Rubus idaeus* is an ideal candidate for genomic library construction since it is diploid ($2n = 2x = 14$) and has a relatively small genome (275 Mbp) (Graham et al. 2004a).

Genomic libraries are stored in vectors that are able to hold large DNA fragments. Normally, yeast artificial chromosomes (YACs), BACs, and genomic libraries require high molecular weight DNA with individual fragments having to be of greater size than 100 kb; therefore, specialized DNA extraction protocols were developed for this purpose (Hein et al. 2005). The first publicly available *Rubus* BAC library was created from "Glen Moy". The library contained over 15,000 clones with an average insert size of 130 kb (almost 8× coverage). Hybridization screening of the BAC library with chloroplast (*rbcL*) and mitochondrial (*nad1*) coded genes revealed that contamination of the genomic library with chloroplast and mitochondrial clones was less than 1% (Hein et al. 2004, 2005). Initial screening of this BAC library employed probes to chalcone synthase, phenylalanine ammonia lyase and a MADS-box gene involved in bud dormancy (I Hein, pers. comm.; B Williamson, pers. comm.). More recently, the library has been probed with genes involved in epidermal cell fate (J Graham, pers. comm.; B Williamson,

Table 3-2 Top 20 *Rubus* species with the largest number of nucleotide and protein sequences in the GenBank April 2010 (*www.ncbi.nlm.nih.gov*).

Species	Nucleotide	EST	Total Sequences	Protein
Rubus alceifolius	8	0	8	0
Rubus arcticus	12	0	12	2
Rubus assamensis	9	0	9	1
Rubus australis	6	0	6	1
Rubus caesius × Rubus idaeus	12	0	12	0
Rubus chamaemorus	7	0	7	1
Rubus corchorifolius	8	0	8	1
Rubus coreanus	10	0	10	4
Rubus crataegifolius	14	0	14	4
Rubus geoides	6	0	6	1
Rubus hochstetterorum	30	0	30	0
Rubus idaeus	225	377	602	114
Rubus longisepalus	6	0	6	0
Rubus parvifolius	22	0	22	5
Rubus phoenicolasius	6	0	6	1
Rubus pungens	7	0	7	1
Rubus rosifolius	6	0	6	3
Rubus saxatilis	6	0	6	1
Rubus trifidus	8	0	8	2
Rubus ulmifolius var. *inermis* × *Rubus ×thyrsiger*	0	2,678	2,678	0
All other taxa	399	0	399	168
	807	3,055	3,862	310

pers. comm.; CE Woodhead, pers. comm.) and a peach ever-growing gene (A Abbott, pers. comm.).

A second library has recently been created from *R. idaeus* cv. Heritage. Unlike the "Glen Moy" library, the inserts are smaller at an average of 70kb, have been inserted into Fosmid vectors, and will be arrayed out in 24,000 clones (almost 7x coverage) (J Swanson, unpubl data). This cultivar was chosen to complement several EST libraries that have been created at the USDA-ARS (Maryland) and at the University of Central Arkansas (K. Lewers, pers. comm. and J Swanson, unpubl data). Fosmids were chosen as they allowed some flexibility in DNA extraction, however more clones need to be arrayed, compared to a BAC library, to represent adequate coverage of the genome.

3.8.3 Expression Libraries and Gene Identification

Expression libraries provide a tool to allow the identification of genes that are expressed in a certain tissue at a particular time. Quite often, clones from these libraries are only partially sequenced to reveal ESTs that are then used in mapping studies or putative gene function is assigned and tested through further experimentation. To date, several expression libraries have been made, most of which are made from raspberry (J Graham, K Smith, M Woodhead and J McCallum, unpubl data; K Lewers, pers. comm.; and Mazzitelli et al. 2007). However, some libraries are now being created from blackberry (Jones and Swanson 2008; Lewers et al. 2008).

Initially expression libraries were used for SSR marker identification (Graham et al. 2004a; Lewers et al. 2008), but they have recently begun to show promise in functional-biology studies (see below). Additionally, subtractive cDNA libraries have recently been used to aid in the identification of genes that may control prickle development (Jones and Swanson 2008). Subtractive libraries enriched for transcripts that are different between two expression libraries. Three subtractive libraries have been created and include a "Heritage" (with thorns) minus "Canby" (almost thornless) from raspberry. A second library was created from siblings showing extreme thorniness minus completely thornlessness of a "Prime Jim" (thorny) × "Arapaho" (thornless) (Fig. 3-1 G-H), and a third library was created from epidermal peels of the "Heritage" minus the cortical tissue of the same individual. To date, over 400 clones of the first library ("Heritage" minus "Canby") have been sequenced, and putative function is currently being assigned.

An additional method for the characterization of genes in *Rubus* is to target genes whose sequence has already been delineated in other species. This already available sequence information is used to create degenerate primers to permit amplification of the the analogous sequence from *Rubus*. These PCR products are often subsequently screened against a cDNA library to obtain the full length gene sequence. This method was used extensively in phylogenetic studies and the first functional studies of *Rubus* genes (Borejsza-Wysocki and Hrazdina 1996; Alice and Campbell 1999; Kumar and Ellis 2001). The success of this approach has been in some debate with between 65% (raspberry to raspberry) to 30% (raspberry to blackberry) specific SSR markers being transferable among *Rubus* species and much less between *Rubus* and *Fragaria* (26–31% in blackberry to strawberry, and 17–21% in raspberry to strawberry) (Lewers and Hokanson 2005b; Stafne et al. 2005). However, in single gene studies, and providing careful attention has paid in degenerate primer creation, it is possible to alter PCR conditions so that amplification is generally successful (J Swanson, pers. obs.). One example of the use of degenerate primer use in red raspberry is by Samuelian et al (2008). Seventy five putative resistance gene analogs were cloned from "Latham" using degenerate primers, eight of which were mapped to a red raspberry map. While functional data is still required to test the validity of these sequences, this study shows the potential utility of this method of gene identification. Although successful, it must be pointed out that when using this method it is only possible to study genes that have been identified in other species. Therefore, it is very likely that unidentified genes that are specific to *Rubus* will be missed unless more random approaches, such as those used in expression libraries, are used.

3.8.4 Functional Studies

Functional genomics focuses on the dynamic aspects of how genes are controlled and how their resulting products (RNA and protein) interact and function. It can be completed at several levels including correlational studies at the transcript or biochemical level, mutational studies where genes or gene products are deleted and the phenotype is observed, and complementational studies where the gene of interest is transformed into a naturally occurring mutant that is deficient in the gene of interest in the hope to recover normal function. The latter two approaches, while being very difficult experimentally, are superior as they show a definite link between the gene and phenotype. The primary studies that have been completed have been involved with fruit ripening, fruit aroma and flavor, disease resistance, and bud dormancy. Functional genomics is, however, still in its infancy in *Rubus*.

3.8.4.1 Fruit Ripening

The identification of genes involved in fruit ripening has been primarily done at the SCRI and has involved cDNA screening approaches as well as RNA-fingerprinting techniques (Jones et al. 1998; Iannetta et al. 2000). These approaches have implicated genes such as pectinmethyl esterase hydrolases (PME) that are involved in cell wall loosening and thus fruit softening, as well as ACC oxidase found in the ethylene biosynthetic pathway (Iannetta et al. 2000). It has long been known that the plant hormone ethylene is important in fruit ripening, and the apparent up-regulation of ACC oxidase seems logical. It was also observed that many of the genes that were up-regulated in the drupelets were also up-regulated in the receptacle indicating that both tissues may be responsive to the same environmental cues thus suggesting that ethylene is a major factor (Iannetta et al. 2000).

3.8.4.2 Phenylpropanoid Pathway and Aroma/Color Properties of Fruit

The end products of the phenylpropanoid pathway contribute significantly to lignin production as well as the color and aroma of *Rubus* fruit (Kumar and Ellis 2001). Members of the phenylalanine ammonia-lyase (PAL) gene family were among the first to be studied. Of the two gene family members of PAL identified, one was found to be differentially expressed (3–10 fold) among various tissues. One gene was associated with early fruit ripening events, while the other was associated with later stages of fruit and flower development indicating differing regulatory mechanisms between the two gene family members (Kumar and Ellis 2001).

The enzymes polyketide synthases (PKS), benzalcetone synthase, and chalcone synthase (CHS) are expressed during fruit development and are partially responsible for the production of the polyketide derivatives benzalacetone, naringenin chalcone and dihydrochalcone that produce some of the color, sweetness, and aroma in the raspberry fruit (Zheng et al. 2001; Kumar and Ellis 2003b; Zheng and Hrazdina 2005, 2008). A number of PKS genes have been characterized from raspberry, and it is suspected that the PKS gene family in *Rubus* consists of at least 11 members; expression analysis showed two of three cDNA studied were up-regulated during fruit ripening (Zheng et al. 2001; Kumar and Ellis 2003b) Additionally, the three studied cDNAs had developmental specific expression patterns, suggesting that a duplication event in *Rubus* gave rise to the independent evolution of regulation and function of these genes (Kumar and Ellis 2003b).

The enzyme 4-coumarate:CoA ligase activates cinnamic acid and its derivatives to thioesters that then serve as substrates for the production of phenylpropanoid-derived compounds that influence fruit quality. Three 4-coumarate:CoA ligase (4CL) genes in raspberry have been identified and are differentially expressed in various organs and during fruit development and ripening (Kumar and Ellis 2003a). Based on the expression patterns and substrate utilization profiles of the recombinant proteins, they found that each of the three genes were expressed in different tissues at different times indicating that the regulation elements may have evolved independently. Furthermore, they suggest that the first gene (*4CL1*) is involved in the phenolic biosynthesis in leaves, the second gene (*4CL2*) is involved in cane lignification, and the third gene (*4CL3*) is involved in the flavonoid and/ or flavor pathway in fruit.

Flavanone 3-hydroxylase (F3H) was recently cloned from black raspberry (Baek et al. 2008). F3H is an enzyme that catalyzes a key step in anthocyanin synthesis and is thought to play a role in the coloration of raspberry fruit. F3H expression was observed throughout fruit development and was greater in the endocarp region of the fruit. Moreover, the anthocyanin content of the fruit was observed to increase 300 fold compared to the green fruit which correlated to the detected increased levels of gene expression of F3H.

3.8.4.3 Phenylpropanoid Pathway and Disease Resistance

The enzyme 4-coumarate:CoA has also been implicated in fungal resistance as it, in conjunction with malonyl-coenzyme A, forms p-hydroxybenzalacetone (Borejsza-Wysocki and Hrazdina 1996). In this case, benzalacetone synthase was shown to carry out part of this reaction: the product of which, p-hydroxybenzalacetone, was shown to inhibit the mycelia growth of the *Rubus* pathogen *P. fragariae* var. *rubi*

Two plant polygalacturonase-inhibiting proteins (PGIPs) have also been cloned from raspberry (Ramanathan et al. 1997). PGIPs inhibit endo-polygalacturonases (endo-PGs) that are released by fungi to help degrade the plant cell wall (Johnston et al. 1993). Of the two genes characterized, one was shown to be nonfunctional due to a frame-shift mutation resulting in truncation of the protein. The second gene was shown to be expressed throughout flower and fruit development (Ramanathan et al. 1997).

3.8.4.4 Bud Dormancy Release

The first microarray experiment in *Rubus* was conducted to investigate bud dormancy phase transition in woody perennial plants at a molecular level (Mazzitelli et al. 2007). Slides were created using a total of 5,300 PCR amplified ESTs from endodormant (true dormancy) and paradormant (apical dominance) raspberry meristematic bud tissue. Over 220 clones were identified that exhibited up or down-regulation during the endodormancy —paradormancy transition. Interestingly, there were a high percentage of genes related to stress tolerance that were identified, and it was attributed that these genes were activating multiple mechanisms to allow the bud to survive low temperatures. Additionally, aquaporins were found to be down-regulated, while cell wall reorganization genes were found to be differentially regulated throughout the time course, and sugar metabolism gene levels also increased. Together, these indicated that water and cell wall reorganization and sugar metabolism were key components of bud dormancy release. Transcription factors, including a SVP-type MADS box transcription factor, and hormone-induced genes were also identified, potentially indicating signaling molecules that may be required to release these buds from dormancy.

3.9 Bioinformatic Resources

As has already been alluded to, the advances in genomics technologies have lead to a massive increase in the numbers of DNA sequences held in public databases. However, Rubus still trails other major rosaceous crops. For example, Rubus (3,862 nucleotide sequences, 310 proteins, April 2010 NCBI search) vastly trails peach (79,567 nucleotide sequences, 310 proteins, April 2010 NCBI search) and other members of the family. However, the number of available *Rubus* sequences is increasing very quickly with over a four-fold increase over the past year, and furthermore, it is much better represented than most rosaceous crops, as most are not cultivated. In order to deal with the increase in genomic and molecular marker data, the Genome Database for Rosaceae (*GDR, http://www.rosaceae.org*) was developed to provide a central bioinformatics resource for researchers of *Rubus*, as well as all Rosaceae species (Jung et al. 2008).

GDR has become one of the central repositories for the current genome sequencing projects that are underway in peach, apple, and strawberry. It currently contains recent assemblies of EST sequencing efforts underway in peach, strawberry, rose, raspberry and blackberry, and apple. Once the complete genome sequences are completed for these species, it will provide an invaluable tool for the *Rubus* community allowing preliminary searches and subsequent degenerate primer design to identify orthologous gene sequences for important genes that have been related to important traits.

3.10 Genomic Resources, the Future

Future work will focus on anchoring the physical map to the genetic map, which will enable alignment of the maps and the identification of genomic regions harboring genes controlling important phenotypes. Some progress has been made here specifically for linkage groups 2, 3 and 6 of the map of Graham et al. (2006) (Graham et al. unrep data). An integrated physical/genetic map will also allow the extent of synteny or colinearity of the *Rubus* genome with other members of the *Rosaceae* to be determined.

With the availability of a detailed genetic linkage map, deep coverage genomic libraries, and several EST libraries we will now be able to identify genetic factors that underpin a wide range of commercially important characteristics. However, the establishment of gene-phenotype relationships will be required. This will require the use of a wide variety of molecular tools including RT-PCR, in situ hybridization, in vitro analysis, and substantial phenotyping. These tools unfortunately will only provide a correlation of gene to trait and it will only be through the implementation of mutational (e.g., RNAi) and complementation studies using transgenic *Rubus* that we can provide solid evidence of gene function in planta.

Once these complex pathways of gene function and their respective regulation are elucidated, we will be able to better use gene-based selection in breeding and the functional assignment of genes for commercially important traits. This in turn will aid breeders allowing for genotype to environment selection so that they may better direct their efforts to create cultivars that are able to maximize yield and fruit quality in a wide variety of specific environments.

3.11 Future Prospects

As we have shown, *Rubus* has been the focus of many breeding projects contributing to significant advances in the hardiness of the plant (including pathogen resistance and environmental tolerance), fruit quality, fruiting time (primocane flowering), thornlessness, and other growth traits (reviewed by (Clark et al. 2007)). However, combining traits together through traditional breeding approaches still remains a difficult task (Clark et al. 2007). Moreover, application of modern molecular approaches in *Rubus* has been hindered by the lack of knowledge regarding the genes and molecular pathways that specify the preferred traits.

As molecular information is accumulated, it will be imperative to have good communication between and among the breeders, molecular researchers, the industry and consumers. This will help best direct the future allocation of resources and prevent unnecessary efforts that may not be in the best interest for the crop in general. In particular, molecular markers may provide the ability to greatly aid breeders by allowing early selection of traits. Traditional plant breeding involves planting thousands of plants each year in specific environments, and 99.0–99.5% of these plants are discarded as they do not have the critically important traits. However, with the aid of molecular markers, critical traits of interest can be selected prior to field planting greatly increasing the potential efficiency of selection in the field. This is one example indicating how the community can work together in a synergistic fashion.

Blackberry and red raspberry production has expanded rapidly over the past two decades and, especially with the general interest in berries and specific interest in foods high in antioxidant levels, this expansion will continue for the foreseeable future. There is a strong push to develop raspberries and blackberries that are better adapted to low chilling and/ or high temperature regions.

Blackberries and raspberries have received a great deal of attention for their potential nutraceutical value. A large number of publications have now established that the dietary intake of berry fruits may have a positive and pronounced impact on human health, performance and disease (Seeram 2008a, b). The International Berry Health Benefits Symposia held in alternate years beginning in 2005 has provided a forum for some of these discussions (Seeram 2008a).

While the interest in nutraceutical value catches a great deal of attention, it cannot be forgotten that caneberries are delightful to eat and good for you. One of the most valuable roles we can play as a Rubus community is to develop cultivars that are more desirable to eat, and can be grown economically by growers so that, in turn, the crop is available at an affordable price for the consuming public.

Acknowledgements

The authors would like to thank John Clark, and Gina Fernandez for their further comments in discussing important traits in *Rubus* and in supplying some of the photographs used (JC). We would also like to thank K. Folta and Tatum Branaman for their kind comments during the preparation of this manuscript.

References

Alice LA, Campbell CS (1999) Phylogeny of *Rubus* (Rosaceae) based on nuclear ribosomal DNA internal transcribed spacer region sequences. Am J Bot 86: 81–97.

Amsellem L, Noyer JL, Le Bourgeois T, Hossaert-McKey M (2000) Comparison of genetic diversity of the invasive weed *Rubus alceifolius* Poir. (Rosaceae) in its native range and in areas of introduction, using amplified fragment length polymorphism Plant Syst Evol 228: 171-179.

Amsellem L, Chevallier MH, Hossaert-McKey M (2001a) Ploidy level of the invasive weed *Rubus alceifolius* Poir (*Rosaceae*), in its native range and in areas of introduction. Plant Syst Evol 228: 171–179.

Amsellem L, Dutech C, Billotte N (2001b) Isolation and characterization of polymorphic microsatellite loci in *Rubus alceifolius* Poir. (Rosaceae), an invasive weed in La Reunion island. Mol Ecol Notes 1: 33–35.

Amsellem L, Noyer JL, Hossaert-McKey M (2001c) Evidence for a switch in the reproductive biology of *Rubus alceifolius* (*Rosaceae*) towards apomixis, between its native range and its area of introduction. Am J Bot 88: 2243–2251.

Anthony VM, Williamson B, Jennings DL, Shattock RC (1986) Inheritance of Resistance to Yellow Rust (*Phragmidium rubi idaei*) in Red Raspberry. Ann Appl Biol 109: 365–374.

Anttonem MJ, Karjalainen RO (2005) Environmental and genetic variation of phenolic compounds in red raspberry. J Food Comp Anal 18: 759–769.

Badjakov I, Todorovska E, Kondakova V, Boicheva R, Atanassov A (2006) Assessment the genetic diversity of Bulgarian raspberry germplasm collected by microsatellite and RAPD markers. J Fruit Ornament Plant Res 14: 61–76.

Baek MH, Chung BY, Kim JH, Wi SG, An BC, Kim JS, Lee SS, Lee IJ (2008) Molecular cloning and characterization of the flavanone-3-hydroxylase gene from Korean black raspberry. J Hort Sci Biotechnol 83(5): 595–602.

Barritt BH, Crandall PC, Bristow PR (1979) Breeding for root rot resistance in red raspberry. J Am Soc Hort Sci 104: 92–94.

Beekwilder J, Hall R, de Vos CHR (2005) Identification and dietary relevance of antioxidants from raspberry. Biofactors 23: 197–205.

Birch ANE, Georghegan IE, Majerus MEN, Hackett C, Allen JM (1997) Interactions between plant resistance genes, pest aphid populations and beneficial aphid predators. SCRI Annual Report for 1996, pp 68–72.

Borejsza-Wysocki W, Hrazdina C (1996) Aromatic Polyketide Synthases: purification, characterization, and antibody development to benzalacetone synthase from Raspberry fruits. Plant Physiol 110: 791–799.

Bristow PR, Daubeny HA, Sjulin TM, Pepin HS, Nestby R, Windom GE (1988) Evaluation of *Rubus* germplasm for reaction to root rot raused by *Phytophthora erythroseptica*. J Am Soc Hort Sci 113: 588–591.

Busemeyer DT, Pelikan S, Kennedy RS, Rogstad SH (1997) Genetic diversity of Philippine *Rubus moluccanus* L. (*Rosaceae*) populations examined with VNTR DNA probes. J Trop Ecol 13: 867–884.

Clark JR (2005) Changing times for eastern United States blackberries. HortTechnology 15: 491–494.

Clark JR, Finn CE (1999) Register of new fruit and nut varieties Brooks and Olmo list no. 39. In: WR Okie (ed) Blackberries and Hybrid Berries. HortScience 34: 183–184.

Clark JR, Finn CE (2002) Blackberry. In: WR Okie (ed) Register of New Fruit and Nut Varieties, list 41. HortScience 37: 251.

Clark JR, Finn C (2006) Blackberry and hybrid berry. Register of new fruit and nut cultivars, list 43. HortScience 41: 1104–1106.

Clark JR, Finn C (2008) Trends in blackberry breeding. Acta Hort 777: 41–48.

Clark JR, Stafne ET, Hall HK, Finn CE (2007) Blackberry breeding and genetics. In: J Janick (ed) Plant Breed Rev 29: 19–144.

Cousineau JC, Donnelly DJ (1989) Identification of raspberry cultivars in vivo and in vitro using isoenzyme analysis. HortScience 24: 490–492.

Crane MB, Lawrence WJC (1931) Inheritance of sex, colour and hairiness in the raspberry, *Rubus idaeus* L. J Genet 24: 243–255

Dale A, McNicol RJ, Moore PP, Sjulin TM (1989) Pedigree analysis of red raspberry. Acta Hort 262: 35–39.

Dale A, Moore PP, McNicol RJ, Sjulin TM, Burmistrov LA (1993) Genetic diversity of red raspberry varieties throughout the world. J Am Soc Hort Sci 118: 119–129.

Darrow GM (1925) The Young dewberry, a new hybrid variety. Am Fruit Grower 45: 33

Darrow GM (1937) Blackberry and raspberry improvement. USDA Yearbook of Agriculture

Daubeny HA (1983) Expansion of genetic resources available to red raspberry breeding programs. Proc 21st Int Hort Congr 1: 150–155.

Daubeny HA (1996) Brambles. John Wiley & Sons, New York, USA.

Daubeny HA (1997) Raspberry, 3rd edn. ASHS Press, Alexandria, VA, USA.

Denoyes-Rothan B, Sasnauskas A, Rugienius R, Chartier P, Petit A, Gordon S, Graham J, Dolan A, Hofer M, Faedi W, Maltoni ML, Baruzzi G, Mezetti B, Sanchez-Sevilla JF, Zurawicz E, Korbin M, Coman M, Mladin P (2008) Genetic resources of European small berries according to GENBERRY project. Sodininkuste IR Darzininkyste 27: 371–377.

Dossett M, Lee J, Finn CE (2008) Inheritance of phenological, vegetative and fruit chemistry traits in black raspberry. J Am Soc Hort Sci 133: 408–417.

Duncan JM, Kennedy DM (1992) Raspberry root rot: A summary of recent progress. SCRI Annu Rep for 1991, pp 89–92.

Espley RV, Hellens RP, Putterill J, Stevenson DE, Kutty-Amma S, Allan AC (2007) Red colouration in apple fruit is due to the activity of the MYB transcription factor, MdMYB10. Plant J 49: 414–427.

Evans KM, Fernandez-Fernandez F, Govan CL (2007) Harmonising fingerprinting protocols to allow comparisons between germplasm collections—*Pyrus*. Proc Eucarpia Fruit Breeding and Genetics Symp, 2007, Zaragoza.

Finn CE (2001) Trailing blackberries: From clear-cuts to your table. HortScience 36: 236–238.

Finn CE (2006) Caneberry breeders in North America. HortScience 41: 22–24.

Finn CE (2008a) *Rubus* spp., blackberry, CABI Publ, Cambridge, MA, USA.

Finn CE (2008b) Temperate fruit crop breeding: Germplasm to genomics. seems incomplete

Finn CE, Knight VH (2002) What's going on in the world of *Rubus* breeding? Acta Hort 585: 31–38.

Finn CE, Hancock JF (2008) Raspberry complete this.

Finn CE, Lawrence FJ, Yorgey B, Strik BC (2001) 'Coho' red raspberry. HortScience 36: 1159–1161.

Finn CE, Swartz HJ, Moore PP, Ballington JR, Kempler C (2002a) Breeders experience with *Rubus* species: *http://www.ars-grin.gov/cor/rubus/rubus.uses.html*

Finn CE, Swartz HJ, Moore PP, Ballington JR, Kempler C (2002b) Use of 58 *Rubus* species in five North American breeding programs-breeders notes. Acta Hort 585: 113–119.

Finn CE, Yorgey B, Strik BC, Martin RR (2005a) 'Metolius' trailing blackberry. HortScience 40: 2189–2191.

Finn CE, Yorgey B, Strik BC, Martin RR, Kempler C (2005b) 'Obsidian' trailing blackberry. HortScience 40: 2185–2188.

Finn CE, Yorgey B, Strik BC, Martin RR, Qian MC (2005c) 'Black Pearl' trailing thornless blackberry. HortScience 40: 2179–2181.

Finn CE, Yorgey B, Strik BC, Martin RR, Qian MC (2005d) 'Nightfall' trailing thornless blackberry. HortScience 40: 2182–2184.

Finn CE, Moore PP, Kempler C (2007) Raspberry cultivars: What's new? What's succeeding? —Where are breeding programs headed? Acta Hort 777: 33–40.

Friedrich A, Váchová J (1999) Transformation of raspberry callus by direct incorporation of plasmid pCB1346 and transfer of this plasmid into apple calluses by cocultivation of these calluses. Acta Hort 484: 587–589.

Galletta GJ, Maas JL, Clark JR, Finn CE (1998) 'Triple Crown' thornless blackberry. Fruit Var J 52: 124–127.

Govan CL, Simpson DW, Johnson AW, Tobutt KR, Sargent DJ (2008) A reliable multiplexed microsatellite set for genotyping *Fragaria* and its use in a survey of 60 *F. ananassa* cultivars. Mol Breed 22: 649–661.

Graham J, McNicol RJ (1995) An examination of the ability of RAPD markers to determine the relationships within and between *Rubus* species. Theor Appl Genet 90: 1128–1132.

Graham J, McNicol RJ, Kumar A (1990) Use of the GUS gene as a selectable marker for Agrobacterium-mediated transformation of *Rubus*. Theor Appl Genet 90: 1128–1132.

Graham J, McNicols R, Greig K, Van de Ven WTG (1994) Identification of red raspberry cultivars and an assessment of their relatedness using fingerprints produced by random primers. J Hort Sci 69: 123.

Graham J, Squire GR, Marshall B, Harrison RE (1997) Spatially dependent genetic diversity within and between colonies of wild raspberry *Rubus idaeus* detected using RAPD markers. Mol Ecol 6: 1001–1008.

Graham J, Smith K, Woodhead M, Russell J (2002) Development and use of simple sequence repeat SSR markers in *Rubus* species. Mol Ecol Notes 2: 250–252.

Graham J, Marshall B, Squire GR (2003) Genetic differentiation over a spatial environmental gradient in wild *Rubus idaeus* populations. New Phytol 157: 667–675.

Graham J, Smith K, MacKenzie K, Jorgenson C, Hackett C, Powell W (2004) The construction of a genetic linkage map of red raspberry (*Rubus idaeus* subsp. *idaeus*) based on AFLPs, genomic-SSR and EST-SSR markers. Theor Appl Genet 109: 740–749.

Graham J, Smith K, Tierney I, MacKenzie K, Hackett C (2006) Mapping gene H controlling cane pubescence in raspberry and its association with resistance to cane botrytis and spur blight, rust and cane spot. Theor Appl Genet 112: 818–831.

Hall HK (1990) Blackberry breeding. Plant Breed Rev 8: 249–312.

Hall HK, Stephens J (1999) Hybrid berries and blackberries in New Zealand—breeding for spinelessness. Acta Hort 505: 65–71.

Hall HK, Cohen D, Skirvin RM (1986a) The inheritance of thornless from tissue culture-derived Thornless Evergreen blackberry. Euphytica 35: 891–898.

Hall HK, Quazi MH, Skirvin RM (1986b) Germplasm release of 'Lincoln Logan', a tissue culture-derived genetic thornless 'Loganberry'. Fruit Var J 40: 134–113.

Hall HK, Quazi MH, Skirvin RM (1986c) Isolation of a pure thornless Loganberry by meristem tip culture. Euphytica 35: 1039–1044.

Hall HK, Stephens MJ, Stanley CJ, Finn CE, Yorgey B (2002) Breeding new 'Boysen' and 'Marion' cultivars. Acta Hort 585: 91–95.

Hall HK, Brewer LR, Langford G, Stanley CJ, Stephens MJ (2003) 'Karaka Black' "Another 'Mammoth" blackberry from crossing eastern and western USA blackberries. Acta Hort 626: 105–110.

Hall HK, Hummer K, Jamieson AR, Jennings SN, Weber CA (2009) Raspberry breeding and genetics. In: Janick J (ed) Plant Breed Rev 32: 39–353.

Haskell G, Garrie JB (1966) Fingerprinting raspberry cultivars by empirical paper chromatography. J Sci Food Agri 17: 189–192.

Hassan MA, Swartz HJ, Inamine G, Mullineaux P (1993) *Agrobacterium tumefaciens*-mediated transformation of several *Rubus* genotypes and recovery of transformed plants. Plant Cell, Tiss Org Cult 33: 9–17.

Hedrick UP (1925) The Small Fruits of New York. J.B. Lyon Albany, NY, USA.

Hein I, Williamson S, Russell J, Graham J, Brennan R, Powell W (2004) Development of genomic resources for red raspberry: BAC library construction, analysis and screening. In: Proc Plant Anim Genome XII Conf, San Diego, CA, USA, p 148.

Hein I, Williamson S, Russell J, Powell W (2005) Isolation of high molecular weight DNA suitable for BAC library construction from woody perennial soft-fruit species. BioTechniques 38.

Holm M, Ma LG, Qu LJ, Deng XW (2002) Two interacting bZIP proteins are direct targets of COP1-mediated control of light-dependent gene expression in *Arabidopsis*. Genes Dev 16: 1247–1259.

Hummer KE, Janick J (2007) *Rubus* iconography: Antiquity to the renaissance. Acta Hort 759: 89–106.

Iannetta PPM, Wyman M, Neelam A, Jones C, Taylor MA, Davies HV, Sexton R (2000) A causal role for ethylene and endo-beta-1,4-glucanase in the abscission of red-raspberry (Rubus idaeus) drupelets. Physiol Plant 110: 535–543.

Jennings DL (1967) Balanced lethals and polymorphism in *Rubus idaeus*. Heredity 22: 465–479.

Jennings DL (1982a) Further Evidence on the Effects of Gene H Which Confers Cane Hairiness on Resistance to Raspberry Diseases. Euphytica 31: 953–956.

Jennings DL (1982b) Resistance to *Didymella applanata* in red raspberry and some related species. Ann Appl Biol 101: 331–338.

Jennings DL (1988) Raspberries and Blackberries: Their Breeding, Diseases and Growth. Academic Press, London, UK.

Jennings DL, McGregor GR (1988) Resistance to cane spot (*Elsinoe veneta*) in the red raspberry and its relationship to yellow rust (*Phragmidium rubi-idaei*). Euphytica 37: 173–180.

Jennings DL, Brydon E (1989) Further studies on breeding for resistance to *Botrytis cinerea* in red raspberry canes. Ann Appl Biol 115: 507–513.

Jennings DL, Daubeny HA, Moore JN (1991) Blackberries and raspberries (*Rubus*). In: JN Moore, J.R. Ballington, (eds) Genetic resources of temperate fruit and nut crops. ISHS, Wageningen,The Netherlands. pp 204–240.

Johnston DJ, Ramanathan V, Williamson B (1993) A protein from immature raspberry fruits which inhibits endopolygalacturonases from *Botrytis cinerea* and other micro-organisms. J Exp Bot 44: 971–976.

Jones CS, Davies HV, McNicol RJ, Taylor MA (1998) Cloning of three genes up-regulated in ripening raspberry fruit (*Rubus idaeus* cv. Glen clova). J Plant Physiol 153: 643–648

Jones N, Swanson J-D (2008) Identification of prickle development genes in Rubus using a subtractive cDNA library. Arkansas Academy of Sciences, Henderson University, AR, USA.

Jung S, Staton M, Lee T, Blenda A, Svancara R, Abbott A, Main D (2008) GDR (Genome Database for Rosaceae): integrated web-database for Rosaceae genomics and genetics data. Nucl Acids Res 36: D1034–D1040.

Kassim A, Poette J, Paterson A, McCallum S, Woodhead M, Smith K, Hackett C, Graham J (2009) Environmental and seasonal influences on red raspberry anthocyanin and antioxidant contents and identification of QTL. Mol Nutri Food Res 53(5): 625–634.

Keep E (1964) Sepaloidy in the red raspberry, *Rubus idaeus* L. Can J Genetic Cytol 6: 52–60.

Keep E (1968) Inheritance of resistance to powdery mildew *Sphaerotheca macularis* (Fr.) Jaczewski in the red raspberry. Euphytica 17: 417–438.

Keep E (1989) Breeding red raspberry for resistance to diseases and pests. Plant Breed Rev 6: 245–321.

Keep E, Knight RL (1967) A new gene from *Rubus occidentalis* L. for resistance to strains 1,2, and 3 of the *Rubus* aphid, *Amphorophora rubi* Kalt. Euphytica 16: 209–214.

Keep E, Knight RL (1968) Use of the black raspberry (*Rubus occidentalis* L.) and other *Rubus* species in breeding red raspberries. Report of East Malling Research Station for 1967, pp 105–107.

Keep E, Knight RL, Parker JH (1970) Further data on resistance to the *Rubus* aphid *Amphorophora rubi* (Kltb.) Report of East Malling Research Station for 1969, pp 129–131.

Kempler C, Daubeny HA, Harding B, Finn CE (2005a) 'Esquimalt' red raspberry. HortScience 40: 2192–2194.

Kempler C, Daubeny HA, Harding B, Kowalenko CG (2005b) 'Cowichan' red raspberry. HortScience 40: 1916–1918.

Kempler C, Daubeny HA, Frey L, Walters T (2006) 'Chemainus' red raspberry. HortScience 41: 1364–1366.

Kempler C, Daubeny HA, Harding B, Baumann CE, Finn CE, Moore PP, Sweeney M, Walters T (2007) 'Saanich' red raspberry. HortScience 42: 176–178.

Knight RL, Keep E (1958) Developments in soft fruit breeding at East Malling. Report of East Malling Research Station for 1957, pp 62–67.

Knight RL, Fernández-Fernández F (2008) Screening for resistance to *Phytophthora fragariae* var. *rubi* in *Rubus* germplasm at East Malling. Acta Hort 777: 353–360.

Knight VH (1993) Review of *Rubus* species used in raspberry breeding at East Malling. Acta Hort 352: 363–371.

Kokko HI, Kärenlampi SO (1998) Transformation of arctic bramble (*Rubus arcticus* L.) by *Agrobacterium tumefaciens*. Plant Cell Rep 17: 822–826.

Kumar A, Ellis BE (2001) The phenylalanine ammonia-lyase gene family in raspberry. Structure, expression, and evolution. Plant Physiol 127: 230–239.

Kumar A, Ellis BE (2003a) 4-Coumarate:CoA ligase gene family in *Rubus idaeus*: cDNA structures, evolution, and expression. Plant Mol Biol 31: 327–340

Kumar A, Ellis BE (2003b) A family of polyketide synthase genes expressed in ripening *Rubus* fruits. Phytochemistry 62: 513–526.

Lawrence FJ (1989) Naming and release of blackberry cultivar 'Waldo'. In: Agr USDoAO (ed) complete this reference.

Lewers KS, Hokanson SC (2005a) Strawberry GenBank-derived and genomic simple sequence repeat (SSR) markers and their utility with strawberry, blackberry, and red and black raspberry. J Am Soc Hort Sci 130: 102–115.

Lewers KS, Hokanson SC (2005b) Strawberry GenBank-derived and genomic simple sequence repeat (SSR) markers and their utility with strawberry, blackberry, and rd and black raspberry. J Am Soc Hort Sci 130: 102–115.

Lewers K, Saski CA, Cuthbertson BJ, Henry DC, Staton ME, Main DE, Dhanaraj AL, Rowland LJ, Tomkins JP (2008) A blackberry (*Rubus* L.) expressed sequence tag library for the development of simple sequence repeat markers. BMC Plant Biol 8.

Lewis D (1939) Genetical studies in cultivated raspberries. I. Inheritance and linkage. J Genet 38: 367–379.

Lewis D (1940) Genetical studies in cultivated raspberries. II. Selective fertilization. Genetics 25: 278–286.

Lindqvist-Kreuze H, Koponen H, Valkonen JPT (2003) Genetic diversity of arctic bramble (*Rubus arcticus* L. ssp. *arcticus*) as measured by amplified fragment length polymorphism. Can J Bot 81: 805–813.

Logan ME (1955) The Loganberry. Mary E. Logan Oakland, CA, USA.

Lopes MS, Maciel B, Mendonca D, Gil FS, Machado AD (2006) Isolation and characterization of simple sequence repeat loci in *Rubus hochstetterorum* and their use in other species from the *Rosaceae* family. Mol Ecol Notes 6: 750–752.

Luo KM, Duan H, Zhao DG, Zheng XL, Deng W, Chen YQ, Stewart CN, McAvoy R, Jiang XN, Wu YH, He AG, Pei Y, Li Y (2007) 'GM-gene-deletor': fused loxP-FRT recognition sequences dramatically improve the efficiency of FLP or CRE recombinase on transgene excision from pollen and seed of tobacco plants. Plant Biotechnol J (5(2): 263–274.

MacFarlane SA, McGavin WJ (2009) Genome activation by *Raspberry bushy dwarf virus* coat protein. J Gen Vir 90: 747–753.

Mallappa C, Yadav V, Negi P, Chattopadhyay S (2006) A basic leucine zipper transcription factor, G-box-binding factor 1, regulates blue light-mediated photomorphogenic growth in *Arabidopsis*. J Biol Chem 281: 22190–22199.

Marking HJ (2006) DNA Marker Analysis of Genetic Diversity in Natural and Cultivated Populations of *Rubus strigosus*, American Red Raspberry. MS Thesis. The Pennsylvania State Univ, University Park, PA, USA.

Marshall B, Harrison RE, Graham J, McNicol JW, Wright G, Squire GR (2001) Spatial trends of phenotypic diversity between colonies of wild raspberry *Rubus idaeus*. New Phytol 151: 671–682.

Martin RR, Mathews H (2001) Engineering resistance to raspberry bushy dwarf virus. Acta Hort 551: 33–37.

Martin RR, Keller KE, Mathews H (2004) Development of resistance to raspberry bushy dwarf virus in 'Meeker' red raspberry. Acta Hort 656: 165–169.

Marulanda ML, Lopez AM, Aguilar SB (2007) Genetic diversity of wild and cultivated *Rubus* species in Colombia using AFLP and SSR markers. Crop Breed Appl Biotechnol 7: 242–252.

Mathews H, Wagoner W, Cohen C, Kellogg J, Bestwick RK (1995) Efficient genetic transformation of red raspberry, *Rubus idaeus* L. Plant Cell Rep 14: 471–476.

Mazzitelli L, Hancock RD, Haupt S, Walker PG, Pont SDA, McNicol J, Cardle L, Morris J, Viola R, Brennan R, Headley PE, Taylor MA (2007) Co-ordinated gene expression during phases of dormancy release in raspberry (*Rubus idaeus* L.) buds. J Exp Bot 58: 1035–1045.

McPheeters KD, Skirvin RM (1983) Histogenic layer manipulation in chimeral 'Thornless Evergreen' trailing blackberry. Euphytica 32: 351–360.

McPheeters KD, Skirvin RM (1989) Somaclonal variation among ex vitro 'Thornless Evergreen' trailing blackberries. Euphytica 42: 155–162.

McPheeters KD, Skirvin RM (2000) 'Everthornless' blackberry. HortScience 35: 778.

Meng R, Chen THH, Finn CE, Li H (2004) Improving in vitro plant regeneration from leaf and petiole explants of 'Marion' blackberry. HortScience 39: 316–320.

Mezzetti B, Landi L, Spena A (2002) Biotechnology for improving Rubus production and quality. Acta Hort 585: 73–78.

Mezzetti B, Costantini F, Chionchetti L, Landi L, Pandolfini T, Spena A (2004) Genetic transformation in strawberry and raspberry for improving plant productivity and fruit quality. Acta Hort .

Moore JN (1984) Blackberry breeding. HortScience 19: 183–185.

Moore JN (1997) Blackberries, 3rd edn. ASHS Press, Alexandria, VA, USA.

Moore PP (1993) Chloroplast DNA diversity in raspberry. HortScience 118: 371–376.

Moore PP (2004) 'Cascade Delight' red raspberry. HortScience 39: 185–187.

Moore PP (2006) 'Cascade Dawn' red raspberry. HortScience 41: 857–859.

Moore PP, Finn CE (2007) 'Cascade Bounty' red raspberry. HortScience 42: 393–396.

Moore PP, Perkins-Veazie P, Weber CA, Howard L (2008) Environmental effect on antioxidant content on ten raspberry cultivars. Acta Hort 777: 499–504.

Morden CW, Weniger DA, Gardner DE (2003) Phylogeny and biogeography of Pacific *Rubus* subgenus *Idaeobatus* (*Rosaceae*) species: Investigating the origin of the endemic Hawaiian raspberry *R. macraci*. Pacific Sci 57: 181 197.

Moyer R, Hummer K, Finn C, Frei B, Wrolstad RE (2002) Anthocyanins, phenolics and antioxidant capacity in diverse small fruits: *Vaccinium*, *Rubus* and *Ribes*. J Agri Food Chem 50: 519–525.

Nybom H, Rogstad SH, Schaal BA (1990) Genetic variation detected by use of the M13 DNA fingerprint in *Malus*, *Prunus* and *Rubus*. Theor Appl Genet 79: 153–153.

Ooka H, Satoh K, Doi K, Nagata T, Otomo Y, Murakami K, Matsubara K, Osato N, Kawai J, Carninci P, Hayashizaki Y, Suzuki K, Kojima K, Takahara Y, Yamamoto K, Kikuchi S (2003) Comprehensive analysis of NAC family genes in *Oryza sativa* and *Arabidopsis thaliana*. DNA Res 10: 239–247.

Ourecky DK (1975) Advances in fruit breeding. Purdue Univ. Press, W. Lafayette, Indiana, USA.

Ourecky DK, Slate GL (1966) Hybrid vigor in *Rubus occidentalis*-R. *leucodermis* seedlings. Proc 17th Int Hort Cong, Abstract 277, vol 1.

Oydvin J (1970) Important breeding lines and cultivars in raspberry breeding. St Forsokag Njos.

Pamfil D, Zimmerman RH, Naess SK, Swartz HJ (1997) Taxonomic relationships in *Rubus* based on RAPD and hybridization analysis (Abst.). HortScience 31: 620.

Pamfil D, Zimmerman RH, Naess K, Swartz HJ (2000) Investigation of *Rubus* breeding anomalies and taxonomy using RAPD analysis. Small Fruits Rev 1: 43–56.

Parent PG, Page D (1992) Identification of raspberry cultivars by non-radioactive DNA fingerprinting. HortScience 27: 1108–1110.

Patamsytė J, Žvingila D, Labokas J, Baliuckas V, Kleizaitė V, Balčiunienė L, Rančelis V (2004) Assessment of diversity of wild raspberries (*Rubus idaeus* L.) in Lithuania. J Fruit Ornament Plant Res 12: 195–206.

Pattison JA, Samuelian SK, Weber CA (2007) Inheritance of phytophthora root rot resistance in red raspberry determined by generation means and molecular linkage analysis. Theor Appl Genet 115: 225–236.

Perkins-Veazie P, Kalt W (2002) Postharvest storage of blackberry fruits does not increase antioxidant levels. Acta Hort 585: 521–524.

Ramanathan V, Simpson CG, Thow G, Lannetta PPM, McNicol RJ, Williamson B (1997) cDNA cloning and expression of polygalacturonase inhibiting proteins (PGIPs) from red raspberry (*Rubus idaeus*). J Exp Bot 48: 1185–1193.

Rusu AR, Pamfil D, Graham J (2006) Mapping resistance of red raspberry (*Rubus idaeus* subsp. *idaeus*) to viral diseases—leaf spot (RLSV) and vein chlorosis (RVCV) on the genetic linkage map. *Buletinul Universitatti de Stiinte Agricole si Medicina Veterinara Cluj-Napoca, Seria Zootechnie si Biotechnologii*, 62: 318–319.

Samuelian SK, Baldo AM, Pattison JA, Weber CA (2008) Isolation and linkage mapping of NBS-LRR resistance gene analogs in red raspberry (*Rubus idaeus* L.) and classification among 270 Rosaceae NBS-LRR genes. Tree Genet Genom 4(4): 881–896.

Sargent DJ, Fernandez-Fernandez F, Rys A, Knight VH, Simpson DW, Tobutt KR (2007) Mapping of A(1) conferring resistance to the aphid *Amphorophora idaei* and dw (dwarfing habit) in red raspberry (*Rubus idaeus* L.) using AFLP and microsatellite markers. BMC Plant Biol 7.

Scott DH, Darrow GM, Ink DP (1957) 'Merton Thornless' as a parent in breeding thornless blackberries. Proc Am Soc Hort Sci 69: 268–277.

Seeram NP (2008a) Berry fruits for cancer prevention: current status and future prospects. J Agri Food Chem 56: 630–635.

Seeram NP (2008b) Berry fruits: Compositional elements, biochemical activities, and the impact of their intake on human health, performance, and disease. J Agri Food Chem 56: 627–629.

Sherman WB, Sharpe RH (1971) Breeding *Rubus* for warm climates. HortScience 6: 147–149.

Siriwoharn T, Wrolstad RE, Finn CE, Pereira CB (2004) Influence of cultivar, maturity and sampling on blackberry (*Rubus* L. hybrids) anthocyanins, polyphenolics, and antioxidant properties. J Agri Food Chem 52: 8021–8030.

Slate GL (1934) The best parents in purple raspberry breeding. Proc Am Soc Hortic Sci 30: 108–112.

Slate GL, Klein LG (1952) Black raspberry breeding. Proc Am Soc Hort Sci 59: 266–268.

Stafne ET, Clark JR, Weber CA, Graham J, Lewers KS (2005) Simple sequence repeat (SSR) markers for genetic mapping of Raspberry and Blackberry. J Am Soc Hort Sci 130: 722–728.

Stanisavljevic M (1999) New small fruit cultivars from Cacak:1. The new blackberry (*Rubus* sp.) cultivar 'Cacanska Bestrna'. Acta Hort 505: 291–295.

Stewart D, McDougall GJ, Sungurtas J, Verrall S, Graham J, Martinussen I (2007) Metabolomic approach to identifying bioactive compounds in berries: Advances toward fruit nutritional enhancement. Mol Nutr Food Res 51: 645–651.

Strik BC (1992) Blackberry cultivars and production trends in the Pacific Northwest. Fruit Var J 46: 207–212.

Strik BC, Clark JR, Finn CE, Bañados P (2007) Worldwide blackberry production. HortTechnology 17: 205–213.

Swartz HJ, Stover EW (1996) Genetic transformation in raspberries and blackberries (Rubus). In: YPS Bajaj (ed) Biotechnology in Agriculture and Forestry. Springer-Verlag, Berlin, 38: 297–307.

Thompson MM (1997) Survey of chromosome numbers in *Rubus* Rosaceae: Rosoideae. Ann MO Bot Gard 84: 128–164.

Trople DD, Moore PP (1999) Taxonomic relationships in *Rubus* based on RAPD analysis. Acta Hort 505: 373–378.

Wada L, Ou B (2002) Antioxidant activity and phenolic content of Oregon caneberries. J Agri Food Chem 50: 3495–3500.

Waldo GF (1948) The Chehalem blackberry. OR Agri Exp Sta Circ 421.

Waldo GF (1950) Notice of naming and release of a new blackberry adapted to the Pacific Coast region. In: Notice USDAR (ed) seems incomplete.

Waldo GF (1957) The Marion blackberry. OR Agri Expt Sta Circ 571.

Waldo GF (1968) Blackberry breeding involving native Pacific Coast parentage. Fruit Var J 22: 3–7.

Waldo GF (1977) Thornless Evergreen-Oregon's leading blackberry. Fruit Var J 31: 26–30

Waldo GF, Wiegand EH (1942) Two new varieties of blackberry the Pacific and the Cascade. OR Agri Exp Sta Circ 269.

Weber CA (2003) Genetic diversity in black raspberry detected by RAPD markers. HortScience 38: 269–272.

Weber CA, Perkins-Veazie P, Moore P, Howard L (2008) Variability of antioxidant content in raspberry germplasm. Acta Hort 777: 493–498.

Williamson B, Jennings DL (1986) Common resistance in red raspberry to *Botrytis cinerea* and *Didymella applanata*, 2 Pathogens Occupying the Same Ecological Niche. Ann Appl Biol 109: 581–593.

Wood GA, Andersen MT, Forster RLS, Braithwaite M, Hall HK (1999) History of Boysenberry and Youngberry in New Zealand in relation to their problems with Boysenberry decline, the association of a fungal pathogen, and possibly a phytoplasma, with this disease. NZ J Crop Hort Sci 27: 281–295.

Woodhead M, McCallum S, Smith K, Cardle S, Mazzitelli L, Graham J (2008) Identification, characterization and mapping of single sequence repeat (SSR) markers from raspberry root rot and bud ESTs. Mol Breed 22(4): 555–563.

Zheng D, Hrazdina G (2005) Expression and function of aromatic polyketide synthases in Rubus. Abstr of Papers of Am Chem Soc 229: U34–U35.

Zheng D, Hrazdina G (2008) Molecular and biochemical characterization of benzalacetone synthase and chalcone synthase genes and their proteins from raspberry (*Rubus idaeus* L.). Arch Biochem Biophys 470: 139–145.

Zheng D, Schroder G, Schroder J, Hrazdina G (2001) Molecular and biochemical characterization of three aromatic polyketide synthase genes from *Rubus idaeus*. Plant Mol Biol 46: 1–15.

4

Strawberry

Part 1: *Fragaria* History and Breeding

Philip J. Stewart

4.1.1 Introduction

No small fruit is enjoyed by more people worldwide than the strawberry, and it stands as one of the most studied of this relatively under-studied group of fruits. The importance of strawberry as a crop, its small size, and phenotypic variability have made it an attractive subject for a number of studies over the years, both for practical studies aimed at improving agricultural outcomes, and for more basic studies aimed at understanding fundamental aspects of plant physiology. In particular, strawberry has been a common subject of experiments focused on the regulation of flowering by environmental cues, for which strawberries exhibit a variety of phenotypes. However, the genetic complexity of the cultivated species, which is octoploid, has lessened the usefulness of the more common varieties of strawberries as a model plant.

4.1.2 Today's Strawberry Industry

The berry crops, due to their perishable nature and frequently brief season, might seem doomed to exist only as minor players in the horticultural world. And yet a combination of humankind's desire for berries, scientific ingenuity, and a remarkably adaptable plant, has made strawberries a major, year-round business. Among the berry crops, they are unrivaled in their economic importance. Worldwide production exceeded 4 million metric tons in 2007, more than double that all other berries combined (FAOSTAT 2008), and the value of the US crop alone was calculated at $1.5 billion (USDA-NASS 2007; http://faostat.fao.org/). In 2007, 263,966 ha were planted

Driscoll Strawberry Associates, Watsonville CA; e-mail: *philip.stewart@driscolls.com* worldwide (FAOSTAT 2008). Leading strawberry producing nations, with their acreage and production, are listed in Table 4.1-1.

Table 4.1-1 Top strawberry producing countries, 2007.

Country	2007 production (tonnes)	Area planted (ha)
United States of America	1,115,000	22,000
Russia	324,000	38,000
Spain	263,900	6,700
Turkey	239,076	10,000
South Korea	200,000	7,000
Japan	193,000	6,800
Poland	168,200	52,500
Mexico	160,000	5,000
Germany	153,000	13,000
Egypt	104,000	3,800
Morocco	100,000	2,500
United Kingdom	66,000	4,000
Ukraine	63,000	11,200
Italy	57,670	2,971
The Netherlands	39,000	2,500
Iran	38,500	3,800
Belarus	33,900	7,400
Serbia	33,129	7,829

Note: Data from FAOSTAT.org, 2008 (some figures are FAO estimates).

Commercial strawberries are grown on every continent but Antarctica and in nearly every country. In many countries, production is mostly on a small scale, using low-input, low-yield methods, for mostly local markets, but in some it has become an intensely managed, large scale endeavor. Such is the case in California, which produces roughly a quarter of the world harvest (USDA-NASS 2007). A mild climate allows for a season as long eight or nine months in some areas, and a long tradition of strawberry breeding has developed a succession of well-adapted cultivars which have allowed California to produce the vast majority of US strawberries for several decades. Other major US production areas capitalize on niches left unsatisfied by the California industry, such as off-season production, as in Florida, or for processing, as in the Pacific Northwest.

Similarly, Spain, Europe's largest producer, has developed its industry to fill the need for off-season fruit in Northern Europe, whose own climate supports only seasonal domestic production outdoors, although greenhouse production is increasingly important in these countries as well. Some countries, such as Russia and Poland, cultivate large areas of strawberry production, but because of low yields these countries have not become proportionately large players in the international market (FAOSTAT 2008). In recent years, the more industrialized nations have begun importing larger quantities of fresh strawberries from nearby developing nations such as Mexico, Morocco, or Egypt.

The traditional method of growing strawberries was as a perennial matted row, allowed to runner freely, protected in the winter, and fruited for several successive seasons. This method is relatively low input, and remains popular in areas where short seasons do not justify expensive inputs. Most open field cultivation in major production areas, however, has moved to the more intensely managed and expensive plasticulture system. In this approach, the plants are planted on raised beds covered with plastic mulch. The plants are fruited for only a single year, and runnering is actively discouraged. Although more expensive, this system has numerous advantages. Disease pressure is reduced due to removal of plant tissue from the field after fruiting and better air circulation, fruit size and cull rates are improved with younger crowns, and the need for plants to survive the off season is eliminated. As a result of changing production systems, the requirements breeders must meet have changed as well.

4.1.3 History

The genus *Fragaria* has likely been a food source for humankind for many thousand years, since our days as hunter-gatherers. As an agricultural crop, however, the strawberry has risen to prominence only in recent centuries, and as result references to it in the ancient literature are few and far between. While major Greek and Roman agricultural and botanical works ignore it completely, Virgil (Clausen 1994) and Ovid (Golding 1567) both briefly mention strawberries in poems of country life, amidst other wild fruits. It was not until over a thousand years later that demand stimulated actual strawberry farming, rather than merely gathering them from wild stands, and as a result the plants went largely unchanged for centuries (Wilhelm and Sagen 1972). Generally collected from diploid species such as *F. vesca* or the hexaploid *F. moschata*, these fruits, though sweet and fragrant, were also extremely small and soft, and production was often sparse and highly seasonal (Darrow 1966).

The introduction of large-fruited New World octoploids, marked the beginning of an upswing in strawberry cultivation. The first to arrive in Europe was *F. virginiana*, from eastern North America. The exact date and means of its arrival are unknown but Wilhelm and Sagen (1972) suggest that it may have been brought by returning colonists from Virginia in 1586, along with tobacco and potatoes. Regardless, it appears to have arrived in Europe by 1629.

The other octoploid arrived nearly a century later. The French army officer, fireworks expert, and spy, Amédée Francois Frezier encountered large-fruited selections of *F. chiloensis* while on a mission to the Spanish colony at Concepción, Chile. He carefully nursed five of these plants, sharing a portion of his own water ration, throughout the six month return journey to France, marking the species' introduction to Europe. As the plants he had chosen were all pistillate, and the need for pollination was poorly understood at the time, the supposedly large-fruited *F. chiloensis* plants were initially a disappointment.

Adoption of these new species, however, was very gradual at first, and octoploid species were little cultivated as late as the middle of the 18th century. By the late 1700s, though, *F. virginiana,* the 'Scarlet' strawberry, had become a popular fixture in English gardens, and new varieties, raised from open-pollinated seed, began to appear, numbering at least 30 by 1820.

The Chilean strawberry, on the other hand, had not garnered much popularity. It was poorly adapted in much of Europe, difficult to grow anywhere but along the coast, prone to winterkill, and unimpressive in its flavor.

One place where *F. chiloensis* did fare relatively well was Brest, on the Atlantic coast of France. Grown alongside *F. virginiana*, the Chilean plants had adequate pollen to set substantial fruit. Although the details are unknown, it may be from these intercrossings in Brest that the earliest *F.* × *ananassa* selections were derived (Darrow 1966).

F. × *ananassa* had appeared in Europe in the middle of the 1700s. Although its origins were uncertain—some said it came from various places in North America, others claimed Suriname or China—its physical appearance, intermediate to the two large-fruited species, led Antoine Duchesne to speculate in 1766 that it was a "cross of the Scarlet strawberry (*F. virginiana*) and the Frutillar (*F. chiloensis*)" and to place it as their descendant on his famous genealogical tree of strawberries (Duchesne 1768). Duchesne, a correspondent of Linnaeus and superintendent of the king's buildings at Versailles, became perhaps the first great strawberry researcher, collecting specimens, producing a systematic and well-reasoned classification of species, sexual compatibility among the different types, and documenting the need for pollination.

The interest in *F. virginiana* in England fired increased attention to the other large-fruited species. British breeder Thomas Knight began what is likely the first true strawberry breeding program in 1817. He continued the intercrossing of the octoploids, producing selections which form the background of many modern cultivars.

Although the bulk of early breeding was done by private entities, often farmers who planted and selected seedlings as a hobby (Hedrick 1925), by the middle of the 20th century much of the emphasis had shifted to public institutions (Darrow 1966). Many US land-grant institutions and state experiment stations have or have had strawberry breeding programs, as have national institutions, like those of the US Department of Agriculture and East Malling and in the UK. Some of these produce cultivars mostly useful for local production, while others have a more global impact. In the latter half of the 20th century, and into this one, private breeding programs have again come to rival public programs in importance. Endeavors by companies like Driscoll Strawberry Associates, Plant Sciences, and California Giant in the US, Edward Vinson in the UK, and Planasa in Spain, have all produced cultivars with worldwide impacts rivaling those of major public breeding programs like the University of California and the University of Florida. Roughly two thirds of 2008 California acreage was planted to proprietary varieties (California Strawberry Commission 2008). Similarly, proprietary cultivars make up at least a fifth of acreage in the Huelva district of Spain (López-Medina and Blanco 2008) and in Florida (Bareuther 2008).

4.1.4 Species

The genus *Fragaria* belongs to the family Rosaceae, home to numerous other treefruits and berry crop species, including raspberries and blackberries, of the genus *Rubus*, with which it shares the subfamily Rosoideae. Its closest relatives are *Duchesnea* Smith and *Potentilla* L. Although the specific geographic origin of the genus is a subject of debate, the rich diversity of species, primarily diploid, in eastern China suggests this area as a possibility (Staudt and Dickorè 2001).

Spread across four continents, strawberries are a complex and varied group of species. The base chromosome number of *Fragaria* is $x = 7$, and species exist across at least four naturally occurring ploidy levels. Most recent authors name roughly 20 species (Hancock and Luby 1993; Folta and Davis 2006). Table 4.1-2 gives the most commonly accepted species and subspecies, along with their ploidy levels. Some of these remain fairly obscure and have been subject to only limited study. Below, some species which have featured most prominently in breeding or genetic studies are discussed briefly.

4.1.4.1 Diploids (2n = 14)

Fragaria vesca is the only species native to both Eurasia and the Americas, and this widespread distribution, along with other factors, has lead some

---

Note for Table 4.1-2

[1]*F. orientalis* may be a tetraploid descendant of *F. mandshurica* (Staudt 1959; 2003).

[2]*F. moupinensis* may be a tetraploid descendant of *F. nilgerrensis* (Darrow, 1966).

[3] *F. nipponica* and *F. yezoensis* share sufficient similarity that some have suggested combining these two taxa.

[4]*F. tibetica* appears to be a tetraploid descendant of *F. pentaphylla* (Staudt and Dickoré, 2001).

Table 4.1-2 *Fragaria* species and subspecies.

Species	Ploidy	Primary Geographic Region of Orgin
F. bucharica	2x	Western Himalayan mountains
F. daltoniana	2x	Eastern Himalayan mountains
F. gracilis	2x	Nortwest China
F. iinumae	2x	Northern Japan, Southern Sakhalin, Eastern Russia
F. mandshurica[1]	2x	Northeastern Asia
F. nilgerrensis[2]	2x	Central Asia
F. nipponica[3]	2x	Japan
F. nubicola	2x	Eastern Himalayan mountains
F. pentaphylla[4]	2x	Chinese Himalayan mountains
F. vesca	2x	North America, Europe, temperate Asia, Hawaii, Chile
ssp. *americana*	2x	Eastern North America
ssp. *bracteata*	2x	Western North America
ssp. *californica*	2x	Coastal California and Oregon
ssp. *vesca*	2x	Europe, temperate Asia
F. viridis	2x	Northern Europe and western Asia
F. yezoensis[3]	2x	Japan
F. × *bifera*	2x/3x	Europe (natural hybrid of *F. vesca* and *F. viridis*)
F. corymbosa	4x	China
F. moupinensis[2]	4x	Southwestern China
F. orientalis[1]	4x	Northeastern Asia
F. tibetica[4]	4x	Eastern Himalayan mountains
F. × *bringhurstii*	5x/6x/9x	Coastal California (natural hybrid of *F. vesca* and *F. chiloensis*)
F. moschata	6x	Northern Europe and parts of western Asia
F. chiloensis	8x	Western North and South America, Hawaii
ssp. *chiloensis*	8x	Coasts and mountains of western South America
ssp. *lucida*	8x	Pacific coast of North America
ssp. *pacifica*	8x	Pacific coast of North America
ssp. *sandwiciensis*	8x	Hawaii
F. iturupensis	8x	Iturup Island (Kuril Islands)
F. virginiana	8x	North America
ssp. *glauca*	8x	Rocky Mountains
ssp. *grayana*	8x	Eastern North America
ssp. platypetala	8x	Rocky Mountains
ssp. *virginiana*	8x	Eastern North America
F. × *ananassa*	8x	Extensively cultivated natural hybrid of *F. chiloensis* and *F. virginiana*
F. × *rosea*	8x+2	Artificial backcross of *Potentilla palustris* into *F.* ×*ananassa*
F. × *vescana*	10x	Artificial hybrids of *F.* ×*ananassa* and *F. vesca*

See note on facing page

to suggest that it originated during the Cretaceous (Staudt 1989), and that a close ancestor may be basal for the genus (Hummer et al. 2008). It exists in a number of subspecies, and is quite phenotypically varied, with documented differences in photoperiodic sensitivity, floral and leaf morphology, fruit and flower color, runnering, crown structure, among other traits (Richardson 1914; Brown and Wareing 1965; Deng and Davis 2001). However, marker studies have shown many individual lines to be quite homozygous (Hadonou et al. 2004).

Because of its ubiquitous nature, small size, self-compatibility, phenotypic diversity, and comparatively simple genomic makeup, *F. vesca* has made an attractive alternative to the more complex cultivated octoploid. In addition to all these factors, at only a little over 200 Mb, the genome of *F. vesca* is only slightly larger than that of *Arabidopsis thaliana* (Folta and Davis 2006), making it an ideal candidate as a model system for Rosaceae (Shulaev et al. 2008). It has also been of interest to breeders because of its flavor and aroma (Trajkovski 1993, 1997).

Native to most of Europe, *F. viridis*, the green strawberry, is notable mainly as the only diploid species other than *F. vesca* to be found outside of Asia. Although Darrow states that it "has no apparent characters that would improve present cultivated sorts, unless it may have a different everbearing or photoperiod response", other authors have noted its tolerance to alkaline soils as a potentially useful trait (Hancock and Luby 1993). Where its habitats overlap with those of *F. vesca*, spontaneous hybrids, designated *F. × bifera*, can occur.

Fragaria nubicola is found in temperate areas of the Himalayas. It is one of several diploid species known to be self-incompatible (Evans and Jones 1967), along with *F. nipponica*, *F. pentaphylla*, and *F. viridis* (Sargent et al. 2004), which *F. nubicola* has been described as resembling (Darrow 1966). Although Staudt (1959) found it to be cross-incompatible with *F. vesca*, later experimentation has given different results (Bors 2000). One particular selection, FDP601, which has been used several times in molecular studies (Sargent et al. 2004b, 2005, 2006), has been the subject of some confusion, and although designated as *F. nubicola* in numerous articles, may in fact be *F. bucharica*, another Himalayan species (Staudt 2006).

Until recently a little known and little studied diploid species, *F. iinumae* has seen an increase in interest after molecular evidence has suggested it as a potential ancestor of the New World octoploids. Although sharing morphological similarities with *F. virginiana* subsp. *glauca*, it has thus far proven incompatible in any crosses with other diploid species (Bors 2000; Sargent et al. 2004) and appears to be evolutionarily isolated from the other diploids (Potter et al. 2000; Rousseau-Gueutin et al. 2009). There appears to be only one report of successful hybridization with *F. iinumae*. Ahmadi and Bringhurst (1992) make two unelaborated references to hybrids with *F. iinumae*, one with *F. virginiana*, the other with complex decaploid hybrids, suggesting that *F. iinumae* may yet prove a resource for genetic improvement of cultivated varieties.

Finally, *F. mandshurica* Staudt is a recently described species (Staudt 2003), which has attracted attention as a possible ancestor of the octoploids (see below). Staudt noted its similarity to *F. orientalis* Losinsk, suggesting that it might be a diploid form of that tetraploid species, whose range it appears to overlap (Staudt 2003; Rousseau-Gueutin et al. 2009).

4.1.4.2 Hexaploids (2n = 42)

There exists only one naturally occurring hexaploid strawberry, *F. moschata*. This species has been in limited cultivation in Europe for roughly 400 years, where it was the largest-fruited strawberry available until the introduction of octoploids from the Americas, but has never been a major crop species. The powerful aroma and musky flavor, while enjoyed by some, is often considered distasteful, and can be overwhelmingly strong in some selections.

The species is normally dioecious, although there do exist perfect flowered cultivars. It has been hypothesized that it is an allopolyploid (Staudt 1959, 2003; Rousseau-Gueutin et al. 2009) with a common ancestor of *F. vesca* as one of the genome consitituents. The involvement of *F. viridis* as chloroplast donor to the hexaploid *F. moschata* is strongly supported by the analyses of the psbJ-psbL and rpl20-rps18 intergenic cpDNA regions (Lin and Davis 2000). It is marginally compatible with *F.* × *ananassa*, and might serve as a source of resistance to diseases such as angular leafspot (*Xanthomonas fragariae*) or powdery mildew (*Sphaerotheca macularis*) (Kantor 1984; Trajkovski 1993; Maas et al. 1995). Darrow (1966) also notes that the fruit tends to ripen simultaneously, something that would be of great value to growers of processing fruit. Much interest in breeding with *F. moschata* has focused on the species' aromatic characteristics. However, Kantor (1984) eventually abandoned attempts to breed octoploid hybrids of *F. moschata* because the characteristic flavors were lost after the repeated backcrosses required to restore the material to the octoploid level.

4.1.4.3 Octoploids (2n = 56)

As mentioned above, the two American octoploid species are the ancestors of the modern cultivated strawberry. *Fragaria virginiana* predominates in North America east of the Rocky Mountains, while *F. chiloensis* is found along the western coasts of both North and South America. The ranges of the two species appear to overlap in the Pacific Northwest, and there appears to be considerable introgression between the two in these areas.

F. virginiana is a highly variable species. It exists as both male and hermaphrodite plants and generally produces pulpy scarlet fruit. These are acidic and aromatic, and have achenes sunk in pits well below the surface. In contrast, the fruit of *F. chiloensis* is larger and firmer, with achenes raised or even with the fruit surface. These are generally less acidic than *F. virginiana*, though often so mild in flavor so as to taste bland. It is a stockier plant, with broad, thick leaves. The species is in general dioecious, with rare hermaphrodites. *Fragaria chiloensis* had been in cultivation by natives along the western edge of South America for many years before the species was discovered by Europeans.

Various wild clones of both species have been noted as having useful characteristics, including: remontant flowering (Hancock et al. 2002), high sugars (Hancock et al. 2003), resistance to *Colletotrichum* species (Lewers et al. 2007), *Verticillium,* and *Phytophthora fragariae* (Maas et al. 1989), low temperatures (Hancock et al. 2003), and drought (Darrow 1966).

It has been suggested that *F. virginiana* and *F. chiloensis* may represent extreme forms of the same species (Hancock et al. 2004). The same work also suggests that many of the current subspecies designations should not be recognized. Currently, *F. virginiana* is generally divided into four subspecies: *virginiana, grayana*, primarily located in eastern North America, and the more western *glauca*, and *platypetala*, while *F. chiloensis* is divided into ssp. *chiloensis, lucida, pacifica,* and the Hawaiian ssp. *sandwicensis.* However, Hancock et al. (2004) question whether *F. chiloensis* ssp. *pacifica* and ssp. *lucida*, and *F. virginiana* ssp. *virginiana* and ssp. *grayana* deserve subspecies rank, and propose the merging of *F. virginiana* ssp. *platypetala* and ssp. *glauca* into a single subspecies. This last group represents strawberries sharing some characteristics of both *F. virginiana* and *F. chiloensis*, located near the intersection of their ranges, and may represent introgression between the two species (Hancock et al. 2004).

4.1.4.4 Natural Decaploids

F. iturupensis, first documented as herbarium samples collected in 1929 (Staudt 1973) has only recently been retrieved as living plants from the island of Iturup, north of Japan (Hummer and Sabitov 2008). As the only Old World octoploid species, it generated some interest, both as a botanical specimen, adding another data point to our understanding of *Fragaria* evolution, and as potential breeding stock. Although it may well serve as the former, it has so far been a disappointment as the latter (Staudt et al. 2008).

Easily the most commercially important species in the genus is the octoploid *F. × ananassa*, whose origins are discussed in the history section above.

4.1.5 Hybridization

As has been previously noted, virtually all cultivated strawberries are *F.* × *ananassa*, a hybrid of *F. chiloensis* and *F. virginiana*. Cultivated strawberry will hybridize freely with either of the two ancestral octoploid species, but crosses with other species are more difficult. Although very little breeding work has yet been conducted with the lesser known octoploid, *F. iturupensis*, the third octoploid species, early studies suggest that the F_1 crosses with *F.* × *ananassa* only with difficulty, and that a high level of sterility is common in the offspring (Staudt et al. 2008).

Although the early European breeders, such as Knight, were limited to those octoploid selections that had been brought to Europe, American breeders had the full range of native germplasm at their disposal. There were numerous introgressions of either *F. virginiana* or *F. chiloensis* in the early years of the crop's development (Darrow 1937; Wilhelm and Sagen 1972). In the early 20th century, private breeders, such as Albert Etter (Fishman 1987) and Luther Burbank (Burbank et al. 1915). made widespread direct use of the native *F. chiloensis*, while C.L. Powers and A.C. Hildreth of the USDA utilized western clones of *F. virginiana* (Hildreth and Powers 1941). Although some of these early crosses have been fundamental to the genetic foundation of the crop, such crosses have made only limited contributions to cultivar breeding in more recent years, and the vast majority of recent breeding has been between *F.* × *ananassa* selections. The germplasm base represented by currently cultivated strawberries is fairly small, calculated to represent no more (and possibly considerably less) than 53 founding clones (of which seven clones account for roughly 50% of current nuclear genes) and 17 cytoplasmic lines (Sjulin and Dale 1987; Dale and Sjulin 1990). This is likely due to a number of factors, including the limited amount of octoploid germplasm available in Europe during the crop's early development, and the need for breeders to continually return to the same few sources of a number of traits, such as everbearing habit, disease resistance, and fruit size or quality.

Any return to wild species for breeding now represents a substantial detriment to many horticultural factors, such as yield, fruit size, and firmness, and most breeders have been reluctant to do so to any great extent. One such occasion represents one of the major developments in strawberry breeding—the introduction of the day-neutrality trait by Royce Bringhurst at the University of California breeding program in the 1970s and 80s (Bringhurst and Voth 1980). This trait was derived from a selection of *F. virginiana* subsp. *glauca*, found in the Wasatch Mountains of Utah, which was noted as flowering under high temperature and long day conditions. This was crossed with advanced selections from the university's breeding program and eventually introduced in cultivars, a process which took four generations (Ahmadi et al. 1990). Today, day-neutral cultivars represent the bulk of California's acreage, although the trait has proved more troublesome to work with in more temperate regions (Dale et al. 2002).

Crosses among many diploid species are possible (Hancock et al. 1991; Sargent et al. 2004), as well as between the hexaploid *F. moschata* and some diploids (Bors and Sullivan 2005) but ploidy and related factors present a substantial barrier to the incorporation of much of the available genetic material into the cultivated octoploid. Interspecific hybridization attempts among *Fragaria* species are summarized in Table 4.1-3. Aside from those originating with the two ancestral octoploid species, only a very few introgressions from other species have made their way out of the experimental stages and into commercial material. These include several decaploid cultivars developed at Balsgård Fruit Breeding Institute in Sweden (Trajkovski 1997) and the Institut for Obstbau der Technischen Universitat Munchen-Weiehnstephan in Germany from crosses with *F. vesca* and dubbed *F. × vescana* (Spiegler et al. 1986). These hybrids appear to contain two genomes from the *vesca* parent in addition to the full octoploid complement of the cultivated parent. Such varieties incorporate

Table 4.1-3 Simply inherited traits described in *Fragaria* species.

Trait	Symbol	Species	Authority
Yellow fruit color	*c*	*vesca*	Williamson et al. 1995;
Seasonal flowering	*SFL (s)*	*vesca*	Richardson, 1914; Brown and Wareing, 1965
Runnerless	*r*	*vesca*	Brown and Wareing 1965
Bushy crown habit	*b*	*vesca*	Brown and Wareing 1965
"Arborea" (Long-stemmed)	*arb*	*vesca*	Guttridge 1973
Pale green leaf	*pg*	*vesca*	Sargent et al. 2004
Pink flowers	*P*	*vesca*	Richardson 1918; Mangelsdorf and East 1927
Resitance to *C. acutatum* pathogenicity group 2	*Rca2*	*x ananassa*	Denoyes-Rothan et al. 2005
Resistance to *P. fragariae*, multiple races	*Rpf1-Rpf5*	*x ananassa*	van de Weg, 1997
Male sterility	*A*	*virginiana*	Spigler et al. 2008
Female fertility	*G*	*virginiana*	Spigler et al. 2008
Sex determination	*Su*	*orientalis*	Staudt 1967
Everbearing	*EV*	*x ananassa*	Monma et al. 1990; Sugimoto et al., 2005
Day-neutrality	-	*x ananassa*	Ahmadi et al. 1990
Appressed petiole hairs	-	*x ananassa, virginiana, chiloensis*	Oydvin 1980; Richardson, 1918
"Monophylla" (Single leaflet)	-	*vesca*	Richardson 1914, 1918

the intense and unusual flavors of the diploid species , but although closer to commercial character than their diploid parents, the small and soft fruits limit their commercial appeal (Trajkovski 2002).

A program at the University of Guelph, in Ontario, Canada, sought to generate synthetic octoploids from lower ploidy material to facilitate the incorporation into cultivar breeding programs (Evans 1977). Two octoploid selections were publicly released from this work, Guelph SO-1, (*F. viridis* x *F. vesca*) x *F. moupinensis* a colchicine-doubled hybrid of *F. moschata* × *F. nubicola* (Evans 1982a) and SO-2 (Evans 1982b). These selections were partially male fertile in crosses with *F.* × *ananassa*, and pollen viability appeared to improve with successive outcrosses (Sangiacomo and Sullivan 1993), but although early reports seemed hopeful that viable varieties derived from these were close at hand (Dale et al 1993) no commercial cultivars appear to have yet been produced from this work.

Although it is relatively difficult to generate healthy plants, intergeneric hybrids with *Potentilla* and *Duchesnea* are possible (Marta et al. 2004). Like the *vescana* hybrids, their commercial utility has thus far been limited. Although seeds and seedlings are readily obtained, the hybrid offspring are generally of poor health and rarely survive long enough to flower. Jelenkovic et al. (1984) found that seedlings from crosses with *P. anserina* mostly died at the cotyledon stage, while those from crosses with *P. fruticosa* consisted either of matroclinous octoploids or hybrid seedlings which grew slowly before dying. Slightly better, though still low, success rates were obtained in crosses between *F.* × *ananassa* and *P. fruticosa* when embryo rescue was used (Sayegh and Hennerty 1993). In crosses between *F. moschata* and *P. fruticosa,* approximately half the seeds germinated, but only nine of 554 survived past the cotyledon stage, and only four of these survived to flower (McFarlane Smith and Jones 1985). In some cases, fertility was restored through colchicine doubling of sterile aneuploid hybrids (Senanayake and Bringhurst 1967).

A handful of ornamental cultivars, with pink flowers derived from *P. palustris*, have been released, notably Lipstick and Pink Panda. These cultivars are the result of multiple backcrosses to *Fragaria*, and at least in the case of the latter are believed to possess a single pair of *Potentilla* chromosomes (Ellis 1991). Such plants have played almost no role in breeding for fruit production, but "Rosalyne" and "Roseberry", from the Quebec breeding program, have 'Pink Panda' as a grandparent and possess pink flowers (Khanizadeh et al. 2002).

So far the contributions of these related genera to breeding are extremely limited, but there are desirable traits which might prove valuable if transferred into strawberries. For example, Delp and Milholland (1981) found *D. indica* remained uninfected after two inoculations with *Colletotrichum fragariae*. The genus *Potentilla* is one of the largest and most varied in the Rosoideae, and may thus harbor considerable genetic potential for strawberry breeders.

4.1.6 Classical Genetics and Traditional Breeding

Compared to many fruit species, strawberries make an attractive target for breeding efforts. They take up little space, allowing for larger seedling populations in a limited area, and the short generation time allows a breeder to cycle through several rounds of selection in the time it would take many tree-fruit breeders to see their seedlings fruit for the first time. These factors, coupled with the potential value of strawberry production, have in part led to the development of dozens of strawberry breeding programs over the years, both public and private, many of which remain active today.

Cultivated strawberry is among the youngest horticultural crops, yet few crop species have been changed so thoroughly, so rapidly. The result has been a steady increase in yields, some of it fueled by cultural advances, but also as a result of ever-improving cultivars which produce larger, firmer fruit in larger and larger amounts. In the past 40 years, average strawberry yield has more than doubled, and in some areas, such as California and Spain, the increase is roughly ten-fold (FAOSTAT data 2007).

4.1.6.1 Breeding for Flowering Habit

Part of this increase in yield, and in the ability of the market to bear the increase in product, has been a result of the development of remontant, or re-blooming cultivars, through the use of day-neutral or everbearing traits. It is believed that the natural state of most strawberry species is as short day photoperiodic plants, initiating flowers primarily in the short days of fall (Darrow 1966). In most temperate regions, strawberries will flower in the spring or early summer, bearing fruit for a matter of weeks, then switch to vegetative growth for much of the remainder of the growing season.

Expression of photoperiodic flowering is heavily influenced by the environment, particularly temperature, with low temperatures favoring flowering under increasingly long day-lengths (Heide 1977; Manakasem and Goodwin 2001). Despite this, there remains a strong genetic component to the response, both in determining the level of photoperiodicity and the required day length for optimal flower initiation. Long season production with short day plants is only possible either in areas where mild winters allow production under short days (such as Florida, Spain, and Mexico), or places where continual cool temperatures can prompt repeat flowering in some of the short day varieties, such as the Watsonville district of California. Even in such locations, short day plants, being more subject to environmental influences, tend to be more erratic in their production from year to year. Remontant strawberries have allowed for more stable production over a longer season, reducing risk and allowing for production in areas and conditions that might not be otherwise viable.

The everbearing character, derived from "Pan-American", probably a sport of the short-day "Bismarck", formed the basis of re-blooming long season cultivars in the Americas for many years, while a similar trait, of uncertain origin, appeared in the European variety "Gloede's Seedling" (Richardson 1914) and possibly other European cultivars (Darrow 1966). However, this material was hampered by a number of weaknesses, which interfered with its adoption into general use. Lack of heat tolerance, poor crown development, poor vigor, and high chilling requirement for years relegated everbearing cultivars to little more than a niche in strawberry production, primarily to allow growers a second season in the fall, as they still do in some areas (Darrow 1966). Although some persisted with this material and eventually overcame many of these problems, most current breeding programs pursuing remontant cultivars have primarily made use of the more recently introduced day-neutrality trait from *F. virginiana* ssp. *glauca*, described previously. Although very similar in many respects to the everbearing trait, the original day-neutral germplasm displayed better heat tolerance, and despite having originated in a wild plant it was quickly introduced into commercial quality cultivars. Cultivars possessing this trait now comprise a majority of the acreage in California, and are grown around the world (Hancock 1999).

4.1.6.2 Other Breeding Progress

Fruits of the wild octoploids rarely average more than a few grams (Hancock et al. 2003). Overall, fruit size has improved markedly in recent decades, although there have always been exceptions in every era, which produced extremely large fruit. It is, however, very difficult to compare fruit size data across studies because most report the average fruit weight of "marketable" fruit, a standard that is both subjective and subject to variation between regions and within a season. Still, the clear trend for commercial varieties, driven by the need for picking efficiency, is ever larger. For example, Darrow (1966) notes the large fruit of "Tioga", a variety released in 1964 by the University of California, which averaged roughly 15 g per fruit (Brooks and Olmo 1972). while "Albion", released by the same program in 2005 and now forming a majority of California production, averages 33 g per fruit according to patent data (Shaw and Larson 2006) and Driscoll Strawberry Associates' "Driscoll Sanibel" averages over 32 g per fruit (Gilford and Mowrey 2006). Some of these gains appear to be the product of increased uniformity in fruit size, eliminating the very smallest and least marketable fruits. However, in some regions, mean fruit size on most local cultivars has remained considerably smaller, often below 15 g (Jamieson et al. 2004; Lewers et al. 2004). These varieties are generally single-cropping short day types, and the need to maximize yield in a single flush of fruit has perhaps favored the selection of highly branching inflorescences with more, rather than larger, fruit. Particularly in the eastern US and Canada, many strawberries are grown for pick-your-own operations, which do not have the same need for harvest efficiency, which has made fruit size a priority in other commercial settings.

It was the introduction of cultivars capable of being shipped long distances by rail, and later truck and air, that triggered the massive growth of large scale strawberry production in many areas. Some of the credit must go to the development of pre-cooling and refrigeration technologies, but it was firm-fruited shipping cultivars like Blakemore and Tennesee Shipper that made strawberries available and relatively affordable for most of the year in most places (Darrow 1966). Unfortunately, while more fruit may today reach the consumer intact than ever before, in many cases less emphasis has been paid by breeders to the flavor and texture of modern strawberry varieties.

Considerable strides have been made to incorporate disease resistance into strawberry cultivars, with varieties resistant to most diseases available. However few diseases have been eliminated as threats. One possible exception is the virus diseases which once routinely devastated strawberries in California. Years of selection have now produced breeding populations, which show general tolerance for these viruses, and while plants may still become infected, it is now rare to see substantial decline as a result. This, combined with new techniques in tissue culture to produce clean stock, has largely eliminated most viruses as a primary concern in strawberries.

Traditional breeding has a long record of success in strawberry and is unlikely to be replaced as the primary means of variety development any time in the near future. That said, there exist opportunities for molecular techniques to contribute to modern breeding efforts. Because the strawberries grown in many areas are the product of generations of breeding for specific conditions and standards, introgression of outside material, even commercial cultivars, often results in substantial setbacks in adaptation or quality. Molecular markers can be used to speed the introgression of characters from outside material and to speed the recovery of desired traits. The need of private programs to move quickly to remain competitive also fuels interest in markers, though few organizations appear to make extensive use of markers in their breeding programs at this time.

4.1.6.3 Inheritance of Simple Traits

Investigations into the genetics of strawberries have been occurring for nearly a century. Richardson, in a series of papers published in 1914, 1918,

1920, and 1923, described the inheritance of a number of traits, in either the diploid *F. vesca* and in octoploids, including runnering, unifoliate leaves, fruit color, and flower color, showing that inheritance in the diploid strawberry conformed to the patterns already documented in other diploid species. Over the years, a number of simply inherited traits have been described (Table 4.1-3) but most have been in *F. vesca*, and while of academic interest have not been of direct utility to breeders.

In the cultivated octoploid, however, relatively few simply inherited traits have been discovered, and most horticulturally important traits appear to be highly quantitative in their inheritance (Table 4.1-4). One prominent set of notable exceptions appears to be disease resistance, for which several

Table 4.1-4 Some studies characterizing phenotypic diversity of useful traits in strawberry species.

Trait	Species	Studies
Resistance to *Colletotrichum* crown rots	*x ananassa* *chiloensis* *virginiana*	Lewers et al. 2007
Resistance to *Xanthomonas fragariae*	*pentaphylla moschata* *nilgerrensis daltoniana* *nubicola gracilis* *iinumae* *vesca* *viridis*	Xue et al. 2005
	x ananassa *chiloensis* *vesca* *virginiana*	Maas et al. 2000
Resistance to *Lygus lineolaris*	*x ananassa* *chiloensis* *virginiana*	Dale et al. 2008
Resistance to *Tetranychus urticae*	*x ananassa* *chiloensis* *virginiana* *x ananassa* *chiloensis* *virginiana* *Potentilla* sp.	Shanks and Moore 1995; Easterbrook and Simpson 1998
Resistance to *Chaetosiphon fragaefolii*	*x ananassa* *chiloensis* *virginiana*	Shanks and Moore 1995
Winter hardiness	*x ananassa* *virginiana* *chiloensis*	Shokaeva 2008; Luby et al. 2007
Day-neutrality	*virginiana*	Hancock et al. 2002
Fruit characteristics	*virginiana chiloensis*	Hancock et al. 2003

major dominant genes are known. van de Weg (1997) described a series of genes conferring race-specific resistance to *Phytophthora fragariae*, the microbial agent responsible for red stele disease. More recently, Denoyes-Rothan et al. (2005) reported on the *Rca2* locus, which appears to confer resistance to pathotype 2 of *Colletotrichum acutatum*, a causal agent of anthracnose in strawberry.

Little progress was made in constructing genetic maps of *Fragaria* using purely phenotypic characters, as most investigations of linkage showed the traits observed to be independently inherited. Brown and Wareing (1965) demonstrated that the runnerless and everbearing traits, suspected by some to be linked because of their coincidence in the alpine type of strawberries, were in fact unlinked recessive traits. One possible case of linkage was demonstrated in that study, however: the bushy crown development habit of the "Bush White" strain of *F. vesca* appeared to be either tightly linked to the locus controlling runner production, or the same locus, with at least three distinct alleles, regulates both runnering and crown development. All three traits were also demonstrated to segregate independently of the fruit color locus. Guttridge (1973) found that the *arborea* locus, conferring a long-stemmed phenotype, also segregated independently of the runnering trait as well. Thus more progress was made in this era in demonstrating the lack of linkages between traits than there was in indentifying linked characters. In fact, no clear instance of genetic linkage in *Fragaria* was reported until the development of isozyme mapping. In 1995 two papers each described separate linked pairs of loci: the PGI-2 isozyme, which segregated with the runnerless locus (Yu and Davis 1995), and SKDH isozyme locus, which was shown to be linked to the yellow fruit color locus (Williamson et al. 1995).

4.1.7 Diversity in *Fragaria*

Despite the wide variety of wild octoploid material available, cultivated strawberries descend from a fairly small group of founding clones. By analyzing pedigrees, Sjulin and Dale (1987) concluded that all cultivated strawberries descend from no more that 53 founding clones, many of which may in fact be related to each other, seven of which likely comprise the sources of 50% of nuclear genes in modern cultivars. A similar follow-up study concluded that among these 53, only 17 have contributed cytoplasms to today's cultivars. In some regions, including California, corelationship between cultivars averaged over 0.4, a figure unlikely to have decreased in the last 20 years. Degani et al (2001) compared amplified fragment length polymorphism (AFLP) and randomly amplified polymorphic DNA (RAPD) markers as measures of diversity with such pedigree-based analyses of diversity.

Because of this narrow pedigree, and the fact that its initial founding wild clones were mostly dictated by what was readily available, not necessarily because it represented the best of such material, Luby et al. (2007) have proposed "reconstructing" *F. × ananassa*, going back to *F. chiloensis* and *F. virginiana* to allow the selection of superior clones. The genetic resources currently available among the wild octoploids for breeding are considerable and still largely untapped. Many of the wild resources remain out of reach to all but the most determined breeders, because of ploidy and sterility barriers, however, and as a result many of these studies have focused their efforts on the octoploids. Even the large and varied populations of the two New World octoploid species remain mostly unexplored by breeders, however, despite evidence that useful sources of important traits may existing among them. Both species offer promise to breeders willing to endure the obstacles posed by breeding with unimproved germplasm.

4.1.7.1 Phylogeny

Currently, the relationships between *Fragaria* species remain relatively poorly resolved by molecular studies, having mostly depended on a small number of loci, which have thus far revealed comparatively low levels of sequence variability.

Studies by Harrison et al. (1997a) and Potter et al. (2000) examined restriction fragment length polymorphism (RFLP) profiles of a chloroplast DNA locus and the ITS (internal transcribed spacer region of rDNA) from a range of strawberry taxa. These studies leave much to be resolved about relationships within the genus, as the loci included in these studies demonstrated only limited variability, the results appear at least to define two distinct clades. The first clade was comprised of the two New World octoploid species, the diploids *F. vesca*, *F. nubicola*, the tetraploid *F. orientalis*, and the hexaploid *F. moschata*. Another group was composed of the Asian diploids *F. daltoniana*, *F. gracilis*, *F. nilgerrensis*, *F. nipponica*, and *F. pentaphylla*. *Fragaria iinumae* appeared to be the most ancestral species, and did not group clearly with either clade. Potter et al. (2000) and Sargent (2004), focused on resolving these relationships among 22 diploid accessions, using five loci (the ITS region and four chloroplast regions). The results once again supported two distinct clades, one containing *F. vesca*, *F. nubicola*, *F. viridis*, and *F. nilgerrensis*, and another with *F. nipponica*, *F. daltoniana*, and *F. pentaphylla*, with *F. iinumae* remaining separated from both. These relationships largely concur with the observations of earlier studies, and are on the whole better supported, but some remain poorly resolved due to a lack of variability among what appears to be a fairly closely related group of species.

An important focus of phylogenetic studies in *Fragaria* has been the origins of the octoploid species. The evidence seems to strongly suggest a common origin for the two New World octoploids, both in molecular studies (Harrison et al. 1997a, b; Potter et al. 2000) and in the high level of fertility in crosses between the two species and their offspring.

Although highly diploidized, these species are alloploids, with a proposed genome composition of AAA'A'BBB'B' (Bringhurst 1990). This leads naturally to the question of where those four distinct genomes originated. *Fragaria vesca* has long been considered a prime candidate as one possible ancestor, beginning with East's (1934) conclusion that one set of chromosomes in *F. virginiana* was sufficiently similar to that of *F. vesca* to allow normal pairing in pentaploid hybrids between the two. In addition to *F. vesca*, several other diploid species have been proposed as possible progenitors. The analysis by Potter et al. (2000) suggested *F. vesca*, *F. nubicola*, and *F. orientalis* as possible donors of both the "A type" genomes and cpDNA. A yet unpublished study by DiMeglio and Davis (cited in Folta and Davis 2006), which examined sequence from two protein-encoding nuclear genes concurs with the suggestions of *F. vesca* and possibly *F. nubicola* in the octoploid's background. This study also proposed *F. iinumae* and *F. mandshurica*, a newly described species (Staudt 2003) not included in previous studies, as possible ancestors. *Fragaria iinumae* has been noted as morphologically similar to *F. virginiana* ssp. *glauca* (Staudt 1999), however no indication of a relationship with the octoploids was observed in the earlier study of the ITS locus (Potter et al. 2000). Although the ITS has been widely used for phylogentic analyses, there is evidence to suggest that region can become homogenized over time in alloploids, potentially eliminating evidence of multiple origins (Wendel et al. 1995).

Another possibility is the elimination of some parts of the original diploid genomes, possibly including the ITS locus. Such genomic alterations have been documented in other species, including extensive studies of newly-synthesized alloploids of *Triticum* and related species (Ozkan et al. 2001), with gene loss occurring as early as the F_1 generation and corresponded to a roughly 6% reduction in C-value (Ozkan et al. 2003). Elimination of specific genes has been documented in numerous other species as well (Wendel 2000), and loss of certain loci in wheat and its relatives appears to be consistent and reproducible (Levy and Feldman 2004).

Correspondingly, the most recent measures of nuclear DNA content in *Fragaria* suggest that the C values for *F. vesca*, estimated at 206 and 211 Mb (Antonius and Ahokus 1996; Akiyama et al. 2001) are noticeably larger than diploid values derived from the values calculated for octoploid species, which ranged from 150 to 200 Mb (Nehra et al. 1991; Nyman and Wallin 1992; Akiyama et al. 2001). Although these may be indicative of differences in genome size in the ancestral diploid species, researchers in other species have noted a slight decrease in mean genome size with increasing ploidy (Tuna et al. 2001) and it appears that such genome downsizing is a very common response to polyploidy in angiosperms (Leitch and Bennett 2004).

Looking forward, breakthroughs in sequencing technologies will greatly facilitate exploration of the strawberry genome. Combined with other technologies dissection of even complicated polyploid genomes may be possible. These new tools may have useful impacts in enabling new breeding strategies that will complement a rich history of traditional efforts.

References

Ahmadi H, Bringhurst RS (1991) Genetics of sex expression in Fragaria species. Am J Bot 78: 504–514.

Ahmadi H, Bringhurst RS (1992) Breeding strawberries at the decaploid level. J Am Soc Hort Sci 117: 856–862.

Ahmadi H, Bringhurst RS, Voth V (1990) Modes of inheritance of photoperiodism in *Fragaria*. J Am Soc Hort Sci 115: 146–152.

Akiyama Y, Yamamoto Y, Ohmido N, Oshima M, and Fukui K (2001) Estimation of the nuclear DNA content of strawberries (*Fragaria* spp.) compared with *Arabidopsis thaliana* by using dual-stem flow cytometry. Cytologia 66: 431–436.

Antonius K, Ahokus H (1996) Flow-cytometric determination of polyploid level in spontaneous clones of strawberries. Hereditas 124: 285.

Bareuther C (2007) The Florida strawberry report. Produce Business 23(12): 39–44.

Bors RH (2000) A streamlined synthetic octoploid system that emphasizes *Fragaria vesca* as a bridge species. PhD Thesis, Faculty of Graduate Studies, Univ of Guelph, Ontario, Canada.

Bors RH, Sullivan JA (2005) Interspecific hybridization of *Fragaria moschata* with two diploid species, *F. nubicola* and *F. viridis*. Euphytica 143: 201–207.

Bringhurst RS, Voth V (1980) Six new strawberry varieties released. Calif Agri 34: 12–15.

Bringhurst RS (1990) Cytogenetics and Evolution in American *Fragaria*. Hortscience 25(8): 879–881.

Brown T, Wareing PF (1965) The genetical control of the everbearing habit and three other characters in varieties of *Fragaria vesca*. Euphytica 14: 97–112.

Burbank L, Whitson J, John R, Williams HS (eds) (1915) Luther Burbank, His Methods and Discoveries and Their Practical Application, vol 6. Luther Burbank Press, New York, London.

Clausen W (1994) Virgil: Eclogues. Clarendon, Oxford Univ Press, Oxford, UK.

Dale A, Sjulin TM (1990) Few cytoplasms contribute to North American strawberry cultivars. HortScience 25: 1341–1342.

Dale A, Daubeny HA, Luffman M, Sullivan JA (1993) Development of Fragaria germplasm in Canada. Acta Hort (ISHS) 348: 75–80.

Dale A, Hancock JF, Luby JJ (2002) Breeding dayneutral strawberries for northern North America. Acta Hort 567: 133–136.

Dale A, Dragan G, Hallet RH (2008) Fragaria virginiana resists tarnished plant bug. Entomol Exp Appl 126: 203–210.

Darrow GM (1937) Strawberry Improvement. USDA Yearbook of Agriculture 1937: 445–495.

Darrow GM (1966) The Strawberry. History, Breeding, and Physiology. Holt, Rinehart and Winston, New York, USA.

Degani C, Rowland LJ, Saunders JA, Hokanson SC, Ogden EL, Golan-Goldhirsch A, Galletta GJ (2001) A comparison of genetic relationship measures in strawberry (*Fragaria x ananassa* Duch.) based on AFLPs, RAPDs, and pedigree data. Euphytica 117: 1–12.

Denoyes-Rothan B, Guérin G, Lerceteau-Köhler E, Risser G (2005) Inheritance of a race-specific resistance to *Colletotrichum acutatum* in *Fragaria* x *ananassa*. Phtyopathology 95: 405–412.

Delp BR, Milholland RD (1981) Susceptibility of strawberry cultivars and related species to *Colletotrichum fragariae*. Plant Dis 65: 421–423.

Deng C, Davis TM (2001) Molecular identification of the yellow fruit color locus in diploid strawberry: a candidate gene approach. Theor Appl Genet 103: 316–322.

Duchesne AN (1768) Histoire Naturelle du Fraisiers, Paris, France.

East EM (1934) A novel type of hybridity in Fragaria. Genetics 19: 167–174.

Easterbrook MA, Simpson DW (1998) Resistance to two-spotted spider mite *Tetranychus urticae* in strawberry cultivars and wild species of *Fragaria* and *Potentilla*. J Hort Sci Biotechnol 73: 531–535.

Ellis JR (1991) Fragaria 'Frel' US patent PP7,598.

Evans WD (1977) The use of synthetic octoploids in strawberry breeding. Euphytica 26: 497–503.

Evans WD (1982a) Guelph SO1 synthetic octoploid strawberry breeding clone. HortScience 17: 833–834.

Evans WD (1982b) Guelph SO2 synthetic octoploid strawberry breeding clone. HortScience 17: 834.

Fishman R (1987) Albert Etter: fruit breeder. Fruit Var J 41: 40–46.

Folta KM, Davis TM (2006) Strawberry genes and genomics. Crit Rev Plant Sci 25: 399–415.

Gilford KL, Mowrey BD (2006) Strawberry plant named 'Driscoll Sanibel' US Patent PP16,298.

Golding, A (1567) The Fifteen Books of P. Ouidius Naso, entytuled Meamorphosis. Willyam Seres, London.

Guttridge CG (1973) Stem elongation and runnering in the mutant strawberry, *Fragaria vesca* L. Arborea Staudt. Euphytica 22: 357–361.

Hadonou AM, Sargent DJ, Wilson F, James CM, Simpson DW (2004) Development of microsatellite markers in *Fragaria*, their use in genetic diversity analysis and their potential for genetic linkage mapping. Genome 47: 429–438.

Hancock JF (1999) Strawberries. CABI Publ, Wallingford, UK.

Hancock JF, Luby JJ (1993) Genetic resources at our doorstep: The wild strawberries. Bioscience 43: 141–147.

Hancock JF, Maas JL, Shanks CH, Breen PJ, Luby JJ (1991) Strawberries (Fragaria) Acta Hort. 290: 491–548.

Hancock JF, Luby JJ, Dale A (1993) Should we reconstitute the strawberry? Acta Hort 348: 86–93.

Hancock JF, Callow PW, Dale A, Luby JJ, Finn CE, Hokanson SC, Hummer KE (2001) From the Andes to the Rockies: Native strawberry collection and utilization. HortScience 36: 206–238.

Hancock JF, Luby JJ, Dale A, Callow PW, Serçe S, El-Sheik A. (2002) Utilizing wild Fragaria virginiana in strawberry cultivar development: Inheritance of photoperiod sensitivity, fruit size, gender, female fertility and disease resistance. Euphytica 126: 177–184.

Hancock JF, Callow PW, Serçe S, Son PQ (2003) Variation in horticultural characteristics of native *Fragaria virginiana* and *F. chiloensis* from North and South America. J Am Soc Hort Sci 128: 201–208.

Hancock JF, Serçe S, Portman CM, Callow PW, Luby JJ (2004) Taxonomic variation among North and South American subspecies of *Fragaria virginiana* Miller and *Fragaria chiloensis* (L.) Miller. Can J Bot 82: 1632–1644.

Harrison RE, Luby JJ, Furnier GR (1997a) Chloroplast DNA restriction fragment variation among strawberry (*Fragaria* ssp.) taxa. J Am Soc Hort Sci 122: 63–68.

Harrison RE, Luby JJ, Furnier GR, Hancock JF (1997b) Morphological and molecular variation among populations of octoploid *Fragaria virginiana* and *F. chiloensis* (Rosaceae) from North America. Am J Bot 84: 612–620.

Hedrick UP (1925) The Small Fruits of New York. State of NY Dept of Farms and Markets, NY, USA.

Heide OA (1977) Photoperiod and temperature intercations in growth and flowering of strawberry. Physiol Plant 40: 21–26.

Hildreth AC, Powers CL (1941) The Rocky Mountain strawberry as a source of hardiness. Proc Am Soc Hort Sci 38: 410–412.

Hummer KE, Sabitov A (2008) Strawberry species of the Iturup and Sakhalin Islands. HortScience 43: 1623–1625.

Hummer K, Bassil N, Davis T, Davidson C, Ellis D, Finn C, Folta K, Hancock J, Höfer M, Luffman M, Martin R, Postman J, Reed B, Retamales J, Roudeillac P, Sjulin T, Tzanetakis Y (2008) Global conservation strategy for Fragaria (strawberry); A consultative document prepared in collaboration with partners in the *Fragaria* germplasm, genetics research-and-development community. Scripta Hort 6.

Jamieson AR, Nickerson NL, Sanderson KR, Privé J-P, Tremblay RJA, Hendrickson P (2004) 'Brunswick' strawberry. HortScience 39: 1781–1782.

Jelenkovic G, Wilson ML, Harding PJ (1984) An evaluation of intergeneric hybridization of Fragaria spp. x Potentilla spp. as a means of haploid production. Euphytica 33: 143–152.

Kantor TS (1984) Results of breeding and genetic work aimed at creating economically valuable cultivars from incongruent crossing of *Fragaria* x *ananassa* Duch. x *Fragaria moschata* Duch. Genetika 19: 1621–1635.

Khanizadeh S, Cousineau J, Deschênes M, Levasseur A, Carisse O (2002) Roseberry and Rosalyne: two new hardy, day-neutral, red flowering strawberry cultivars. Acta Hort (ISHS) 567: 173–174.

Leitch IJ, Bennet MD (2004) Genome downsizing in polyploid plants. Biol J Linn Soc 82: 651–663.

Levy AA, Feldman M (2004) Genetic and epigenetic reprogramming of the wheat genome upon allopolyploidization. Biol J Linn Soc 82: 607–613.

Lewers KS, Enns JM, Wang SY, Maas JL, Galletta GJ, Hokanson SC, Clark JR, Demchak K, Funt RC, Garrison SA, Jelenkovic GL, Nonnecke GR, Probasco PR, Smith BJ, Smith BR, Weber CA (2004) 'Ovation' strawberry. HortScience 39: 1785–1788.

Lewers KS, Turechek W, Hokanson SC, Maas JL, Hancock JF, Serçe S, Smith BJ (2007) Evaluation of elite native strawberry germplasm for resistance to anthracnose crown rot disease caused by *Colletotrichum* sp. J Am Soc Hort Sci 132: 842–849.

Lin J, Davis TM (2000) S1 analysis of long PCR heteroduplexes; detection of chloroplast indel polymorphisms in Fragaria. Theor Appl Genet 101: 415–420.

López-Medina J, Blanco C (2008) Small fruit production systems in south western Spain: Huelva. Euroberry COST 863 Small Fruits Production Systems Working Group, Pula, Croatia, 29–31 May, 2008 (Abstract).

Luby JJ, Hancock JF, Dale A, Serçe S (2007) Reconstructing *Fragaria* x *ananassa* utilizing wild *F. virginiana* and *F. chiloensis*: inheritance of winter injury, photoperiod sensitivity, fruit size, female fertility and disease resistance in hybrid progenies. Euphytica 163: 57–65.

Maas JL, Galletta GJ, Draper AD (1989) Resistance in strawberry to races of *Phytophthora fragariae* and to isolates of *Verticillium* from North America. Acta Hort 265: 521–526.

Maas JL, Pooler MR, Galletta GJ (1995) Bacterial angular leafspot disease of strawberry: Present status and prospects for control. Adv Strawberry Res 14: 18–24.

Maas JL, Gouin-Behe C, Hartung JS, Hokanson SC (2000) Sources of resistance for two differentially pathogenic strains of *Xanthomonas fragariae* in *Fragaria* genotypes. HortScience 35: 128–131.

Manakasem Y, Goodwin PB (2001) Responses of day neutral and June bearing strawberries to temperature and daylength. J Hort Sci Biotechnol 76: 629–635.

Mangelsdorf AJ, East EM (1927) Studies on the genetics of *Fragaria*. Genetics 12: 307–339.

Marta AE, Camadro EL, Diaz-Ricci JC, Castagnaro AP (2004) Breeding barriers between the cultivated strawberry *Fragaria x ananassa*, and related wild germplasm. Euphytica 136: 139–150.

McFarlane Smith WH, Jones JK (1985) Intergeneric crosses with *Fragaria* and *Potentilla*. I. Crosses between *Fragaria moschata* and *Potentilla fruticosa*. Euphytica 34: 725–735.

Monma S, Okitsu S, Takada K (1990) Inheritance of the everbearing habit in strawberry. Bull. Natl. Res. Inst. Veg. Ornam. Plants Tea Jpn Ser C 1: 21–30.

Nehra NS, Kartha KK, Stushnoff C (1991) Nuclear DNA content and isozyme variation in relation to morphogenic potential of strawberry (*Fragaria x ananassa*) callus cultures. Can J Bot Rev 69: 239–244.

Nyman M, Wallin A (1992) Improved culture technique for strawberry (*Fragaria x ananassa* Duch) protoplasts and the determination of DNA content in protoplast derived plants. Plant Cell Tiss Org Cult 30: 127–133.

Oydvin J (1980) Inheritance of type of pubescence on the pedicels of strawberries (*Fragaria x ananassa* Duchesne). Meldinger fra Norges Landbrukshogskole 59: 1–9.

Ozkhan H, Levy AA, Feldman M (2001) Allopolyploidy-induced rapid genome evolution in wheat (Aegilops-Triticum) group. Plant Cell 13: 1735–1747.

Ozkhan H, Tuna M, Arumuganthan K (2003) Nonadditive changes in genome size during allopolypodization in wheat (*Aegilops-Triticum* group). J Hered 94: 260–264.

Potter D, Luby JJ, Harrison RE (2000) Phylogenetic relationships among species of *Fragaria* (Rosaceae) inferred from non-coding nuclear and chloroplast DNA sequences. Syst Bot 25: 337–348.

Richardson CW (1914) A preliminary note on the genetics of Fragaria. J Genet 3: 171–177.

Richardson CW (1918) A further note on the genetics of *Fragaria*. Journal of Genetics 7: 167–170.

Richardson CW (1920) Some notes on *Fragaria*. Journal of Genetics 10: 39–46.

Richardson CW (1923) Notes on *Fragaria*. Journal of Genetics 13: 147–152.

Rousseau-Gueutin M, Gaston A, Aïnouche AK, Aïnouche ML, Olbricht K, Staudt G, Richard L, Denoyes-Rothan B (2009) Tracking the evolutionary history of polyploidy in *Fragaria* L. (strawberry): New insights from phylogenetic analyses of low-copy nuclear genes. Mol Phylogen and Evol 51: 515–530.

Sangiacomo MA, Sullivan JA (1993) Introgression of wild species into the cultivated strawberry using synthetic octoploids. Theor Appl Genet 88: 349–354.

Sargent DJ, Geibel M, Hawkins JA, Wilkinson MJ, Battey NH, Simpson DW (2004) Quantitative and qualitative differences in morphological traits revealed between diploid *Fragaria* species. Ann Bot 94: 787–796.

Sayegh AJ, Hennerty MJ (1993) Intergeneric hybrids of Fragaria and Potentilla. Acta Hort 348: 151–154.

Shanks CH, Jr., Moore PP (1995) Resistance to two spotted spider mite and strawberry aphid in *Fragaria chiloensis, F. virginiana,* and *F. x ananassa* clones. HortScience 30: 596–599.

Shaw DV, Larson KD (2006) Strawberry plant named 'Albion' US Patent PP16,228.

Shokaeva DB (2008) Injuries induced in different strawberry genotypes by winter freeze and their effect on subsequent yield. Plant Breed 127: 197–202.

Shulaev V, Korban SS, Sosinski B, Abbot AG, Aldwinckle HS, Folta KM, Iezzoni A, Main D, Arús P, Dandekar AM, Lewers K, Brown SK, Davis TM, Gardiner SE, Potter D, Veilleux RE (2008) Multiple models for Rosaceae genomics. Plant Physiol 147: 985–1003.

Sjulin TM, Dale A (1987) Genetic diversity of North American strawberry cultivars. J Am Soc Hort Sci 112: 375–385.

Spiegler G, Schlindwein B, Schimmelpfeng H (1986) Untersuchungen zur Selektion von deaploiden *Fragaria x vescana*. Erwerbsobstbau 28: 220–221.

Staudt G (1959) Cytotaxonomy and phytogenetic relationships in the genus *Fragaria*. Proc 9th Int Bot Congr Montreal 2: 377.

Staudt G (1973) *Fragaria iturupensis*, eine neue Erdbeerart aus Ostasien. Willdenowia 7: 101–104.

Staudt G (1989) The species of *Fragaria*, their taxonomy and geographic distribution. Acta Hort 265: 23–33.

Staudt G (2003) Notes on Asiatic *Fragaria* species: III. *Fragaria orientalis* Losinsk. and *Fragaria mandhurica* spec. Nov Bot Jahrb 124(4): 397–401.

Staudt G (2006) Himalayan species of *Fragaria* (Rosaceae). Botanische Jahrbücher 126: 483–508.

Staudt GS, Dickoré WB (2001) Notes on Asiatic Fragaria species: Fragaria pentaphylla Losinsk. and Fragaria tibetica spec. nov Bot Jahrb Syst 123: 341–354.

Staudt G, Schneider, Scheewe P, Ulrich D, Olbricht K (2008) *Fragaria iturupensis*, a new source for strawberry improvement? VI Int Strawberry Symp, Huelva, Spain, March 3–7, 2008, p 49 (abstract).

Strobel JW (1967) Superior new strawberry clones resist Verticillium wilt. Proc Fla Soc Hort Sci 80: 138–143.

Sugimoto T, Tamaki K, Matsumoto J, Yamamoto Y, Shiwaku K, Watanabe K (2005) Detection of RAPD markers linked to the everbearing gene in Japanese cultivated strawberry. Plant Breed 124: 498–501.

Tuna M, Vogel KP, Arumuganathan K, Gill KS (2001) DNA content and ploidy determination of bromegrass germplasm accessions by flow cytometry. Crop Science 41(5): 1629–1634.

Trajkovski K (1993) Progress report in *Fragaria* species hybridization at Balsgård, Sweden. Acta Hort 348: 131–136.

Trajkovski K (1997) Further work on species hybridization in Fragaria at Balsgård. Acta Hort 439: 67–74.

Trajkovski K (2002) Rebecka, a day-neutral *Fragaria x vescana* variety from Balsgård. Acta Hort 567: 177–178.

USDA-NASS (2007) Statistical highlights of U.S. Agriculture 2006 & 2007: *http://www.nass.usda.gov/Publications/Statistical_Highlights/2007/2007STATHI.PDF* (accessed 2 Aug 2008).

van de Weg WE (1997) A gene-for-gene model to explain interactions between cultivars and races of *Phytophthora fragariae* var. *fragariae*. Theor Appl Genet 94: 445–451.

Wendel JF (2000) Genome evolution in polyploids. Plant Mol Biol 42: 225–249.

Wendel JF, Schnabel A, Seelman T (1995) Bidirectional interlocus concerted evolution following allopolyploid speciation in cotton (Gossypium). Proc Natl Acad Sci USA 92: 280–284.

Wilhelm S, Sagen JE (1972) A history of the strawberry from ancient gardens to modern markets. Agricultural publ, Univ of California, Berkeley, USA.

Williamson SC, Yu H, Davis TM (1995) Shikimate dehydrogenase allozymes: Inheritance and close linkage to fruit color in the diploid strawberry. J Hered 86: 74–76.

Xue S, Bors RH Streikov SE (2005) Resistance sources to *Xanthomonas fragariae* in non-octoploid strawberry species. HortScience 40: 1653–1656.

Yu HR, Davis TM (1995) Genetic-linkage between runnering and phosphoglucoisomerase allozymes, and systematic distortion of monogenic segregation ratios in diploid strawberry. J Am Soc Hortic Sci 120(4): 687–690.

Strawberry

Part 2: Genome Composition, Linkage Maps and Markers

Kevin M. Folta,[1]* Béatrice Denoyes-Rothan,[2] Mathieu Rousseau-Gueutin[2] and Philip J. Stewart[3]

4.2.1 Introduction

If you want to begin to understand the genetic composition of strawberry first hand, a good place to start is the USDA Germplasm Repository in Corvallis, Oregon, USA. Here strawberry plants exist that have been collected from locales all over the globe, representing the *Fragaria* genus at all levels of its ranging variation. Inside one of the many greenhouses you can truly appreciate the genetic complexity resident within the genus *Fragaria*. The inherent variation in what is normally considered to be a rather monomorphic plant stands out when the individuals can be examined side by side. You can easily detect conspicuous differences in plant stature, leaf size, flower morphology and fruit size, along with many other attributes that only increase in number the more you observe them. These differences reflect the broad adaptation that has occurred among *Fragaria* species as they spread and colonized discrete niches throughout the northern hemisphere and the west coast of South America (for review see Darrow 1966; Hancock 1999; Hummer and Hancock 2009; Staudt 2009).

The same natural variation would inspire human intervention to affect the radiation of these species and eventual use of this plant in cultivation. While domestication of strawberry has been known in Europe (mainly for

[1]Department of Horticultural Sciences and the Graduate Program in Plant Molecular and Cellular Biology, University of Florida, Gainesville, FL 32611, USA.
[2]INRA, UR 419, Unité de Recherche sur les Espèces Fruitières, Domaine de la Grande Ferrade, Villenave d'Ornon, France.
[3]Driscoll's Strawberry Associates, Watsonville, CA, 95077 USA.
*Corresponding author: *kfolta@ufl.edu*

F. vesca and *F. moschata* and some *F. virginiana*) and Chile (*F. chiloensis*) for centuries, one of the greatest stories is the tremendous work by a relatively uncelebrated French scientist and artist, A. E. Duchesne. While studying plants as a teenager in Versailles, Duchesne performed crosses, defined fertility groups and noted the inheritance of phenotypes 100 years before Mendel and his peas. Some of the same atypical plants described then, such as the single-leaflet bearing *F. vesca* subsp. *vesca* (syn. *F. monophylla Duch.*), are still puzzling today. Duchesne made incredibly untimely inferences about the relationships and lineage between various strawberry accessions, ranking them from ancestral and derived forms. He laid a foundation of evolutionary studies almost a century before Darwin, and spawned questions in strawberry evolution that are only now being formally answered. Contemporary strawberry genetics and genomics represent the tail end of a surprisingly ancient science.

4.2.2 Many Levels of Ploidy

If you walk through the doors of a greenhouse at the strawberry germplasm repository one can look straight down an aisle that separates two sets of very typical greenhouse benches, one to the left and one to the right (Fig. 4.2-1). On the right are the diploids. On the left, the octoploids, plants carrying close to four complete genomes. Polyploidy, which is the occurrence of more than two copies of the basic set of chromosomes in a somatic cell, is prevalent in plants. At least 70% of angiosperms have experienced one or more events of genome duplication (complete or partial) in their evolutionary history (Bowers et al. 2003). The prominence of polyploidy in higher plants suggests that it is a key factor in plant evolution leading to the formation of new species (Cui et al. 2006). The effect of polyploidization is immediately obvious, as the plants to the left teem with vegetation.

Figure 4.2-1 An example of the effect of polyploidization in strawberry at the USDA Germplasm Repository in Corvallis, Oregon, USA. The bench on the right contains diploid strawberry species, the bench on the left shows octoploids. The significant differences in stature and vigor are conspicuous in plants in the foreground.

This process affects nucleus, cell and organ size that become larger with increasing levels of ploidy, and also the period of meiosis and mitosis, which is longer in polyploids. The polyploid strawberry plants at the germplasm repository are tall, full, and even the untrained eye would note these as model portraits of preferred plant vigor. Another consequence of polyploidy is the presence of multiple alleles for one gene. This process leads to higher genetic diversity and fixed heterozygosity at duplicated loci (Comai 2005; Wendel 2000). Polyploidy is a large part of the strawberry's commercial success. Clearly polyploidization has played a central role in strawberry evolution and domestication, as the major cultivars today are invariably octoploids. The clear difference in plants of varying ploidy offers many basic questions and practical applications.

Observation of the clear effects of polyploidization initiates a new series of important questions. Which ancestrol diploid genomes are present as subgenomes of the polyploid? How do they segregate? How do they contribute to important plant traits? Strawberry structural genomics took a major step forward with the recent sequencing of the diploid strawberry *F. vesca*. But even the complete genome sequence does not permit an easy resolution of ancestral relationships between species within the genus. For this, researchers have relied on a broad series of approaches that started over 80 years ago, ranging from cytological to molecular. These methods have presented substantial information about genome structure.

Fragaria chromosomes became a subject of study in 1921 at the Bussey Institution at Harvard University. Here Ichijima performed cytological evaluation of *Fragaria* materials provided by George M. Darrow, a USDA scientist who does not need an introduction among small fruit researchers. Cytological studies began in earnest in 1924, focusing on a series of diploids and some interesting hybrids. Analysis of the chromosomes in pollen indicated a haploid chromosome complement of seven. Analysis of somatic cells from diploid and octoploid species detailed the structure and behaviors of chromosomes in great detail. The study also analyzed chromosome pairing in hybrids. One hybrid was particularly revealing, the cross between *F. vesca bracteata* ($x = 7$) and *F. virginiana* ($x = 28$). The F_1 progeny possessed seven bivalents during meiosis and 21 univalents, illustrating that the diploid chromosomes could pair successfully with those in an octoploid and successfully negotiate meiosis, while non-bivalents moved randomly on the spindle (Ichijima 1926). These findings illustrate that the diploid chromosomes pair correctly with a subset of those resident within the octoploid nucleus, supporting the hypothesis that diploid and octoploid genomes are similar. This hypothesis was recently confirmed using comparative mapping (Rousseau-Gueutin et al. 2008). Ichijima's groundbreaking work was delayed by illness, but in 1926 was finally published (Ichijima 1926). His work emerged in the literature contemporaneously with an independent study of meiotic chromosomes by Longley (1926) that agreed well with the findings, dividing *Fragaria* into three basic groups. These were the diploids, hexaploids and octoploids, again with a basic chromosome number of seven. Later cytological studies were built on these original observations. The chromosomes of *Fragaria* are small, yet 14 metaphase chromosomes could be clearly resolved in diploid species (Iwatsubo and Naruhashi 1989). Counting chromosomes in octoploid and decaploid species is difficult because the chromosomes are densely positioned, yet recent reports clearly present an accounting of all chromosomes.

There is a great natural variation in ploidy. This basic chromosome complement is the foundation of various higher-order multiples that define at least four (and possibly five) levels of natural polyploidy. The diploids are the fundamental unit of the genome that have pyramided themselves through gametic non-reduction, outcrossing, and duplication to give rise to the modern day polyploids. The higher levels of ploidy are essentially higher order organizations of one or more diploid genomes hosted by a single polyploid nucleus. Unreduced gametes are common in strawberry (Dickinson et al. 2007), as approximately 1% of pollen grains are enlarged (Bringhurst and Senanayake 1966). There is no evidence that indicates if modern polyploids arose from single polyploidizations and speciation or multiple independent events that underlie a given subset of polyploid wild species, but these relationships are starting to be elucidated, and will be discussed at the end of this chapter.

Diploid strawberry ($2n = 2x = 14$) is found naturally throughout the northern hemisphere and is cultivated in some areas of Europe in a few commercial niches. Bors and Sullivan (1998) define three major interfertile groups. The first is comprised of *F. vesca*, *F. pentaphylla*, *F. viridis* and *F. bucharica*. The second contains again, *F. vesca*, and also *F. nilgerrensis*, *F. daltoniana* and *F. pentaphylla*. The third contains *F. gracilis* Losinsk, *F. nipponica* and *F. pentaphylla*. The Japanese diploid *F. iinumae* has not been sufficiently tested for interfertility, but it does not successfully cross with *F. vesca*, *F. bucharica* Losinsk or *F. viridis* Weston.

The dominant radiation is *F. vesca*, which is found throughout Europe and North America. It is the only diploid in North America, comprised of subspecies *F. vesca americana* and *F. vesca bracteata*. Another diploid, *F. viridis*, has been reported to maintain an exceedingly small genome, on par with *Arabidopsis thaliana* (Hodgson 2007). Recent evidence has brought great attention to *F. iinumae*, a Japanese diploid that appears to be a central donor to the modern cultivated strawberry (Rousseau-Gueutin et al. 2009). *F. iinumae* is unique in several ways. Its chloroplast restriction fragment length polymorphism (RFLP) patterns are distinct from most other diploid strawberries (Harrison et al. 1997), it has a glaucous leaf phenotype that is distinctive, and it appears extremely susceptible to powdery mildew and other pathogens. Other diploids round out the category, including *F. viridis*.

Tetraploid species ($2n = 4x = 28$) are found throughout Asia, generally at the edge of diploid ranges. In general the tetraploids are considered to be autopolyploidization of diploids, and are represented by the species *F. orientalis, F. graciliis, F. tibetica* Staudt & Dickoré and *F. moupinensis* (Franch.) Cardot. Analyses of low copy nuclear genes showed that these species were distributed into two clades, which suggests at least two independent events of polyploidization. The first clade included only the tetraploid *F. orientalis*, which shows a typical Eurasian-American distribution, and the two diploid species *F. vesca* and *F. mandshurica*, which is consistent with previous results from cpDNA and rDNA analyses (Potter et al. 2000). In addition, contribution of *F. mandshurica* as one of the parent donors of *F. orientalis* is strengthened by the evidence of overlapping geographical distribution (eastern Asia, Mandchuria and Korea) between these two species (Staudt 1989, 2003). The second clade was composed of the other four tetraploid species, which are distributed today in East and Southeast Asia, and grouped also the diploid species *F. nipponica, F. nubicola, F. pentaphylla* and *F. yezoensis*. Using these nuclear genes, no clear conclusion can be drawn on the possible subgenome contributions from the diploids to these tetraploids.

The higher-ploids include the octoploids ($2n = 8x = 56$). As mentioned previously, the precise constituents of the octoploid genome remained a topic of discussion and inquiry, perhaps even debate. As stated in the previous chapter, two principle wild species *F. virginiana* and *F. chiloensis* represent the immediate progenitors that later hybridized to produce the cultivated octoploid, *F.* × *ananassa*. Their distributions originated in the New World with *F. virginiana* populating the understories and fields of North America, while *F. chiloensis* inhabits the west coast of the Americas to this day.

A wild decaploid has also been characterized (Hummer et al. 2009). *F. iturupensis* Staudt was identified from a 1929 Japanese herbarium specimen that originated from Atsunupuri volcano on the island of Iturup north of Japan. Living specimens were gathered at the turn of this century, and analysis of kayotypes indicated that this unusual plant carried a decaploid chromosome complement (Hummer et al. 2009) while octoploid numbers of chromosomes have been previously reported (Staudt, 1973). The identification and characterization of *F. iturupensis* reminds us that the occurrence of polyploidy in strawberry likely does not follow a simple and predictable lineage, as the modern plants may derive from many independent polyploidization events that could involve a cross-section of diploid subgenome constituents. While a formal topic of scientific inquiry for almost 100 years, the benefit of molecular tools now is illuminating the subgenome composition of polyploid strawberry.

4.2.3 Genome Composition

The strawberry genome is a miniscule plant genome. A quantitative approach to genome size is addressed in the next chapter, so skim coverage of the topic is presented here. The total genome size has been estimated for five strawberry species, with a value of approximately 200 Mb (reviewed in Folta and Davis 2006). This value is based on original estimates by Akiyama (2001) that were based on an *Arabidopsis* reference that would change so the 163 Mb estimate eventually needed to be rescaled as the *Arabidopsis* values were updated (Bennett et al. 2003). Curiously, the 1x C values for octoploid strawberry are significantly lower than those for diploid. Estimates by several independent groups have produced corrected haploid genome sizes from the octoploid strawberry at 155, 177, 180 and 200 Mb (Nehra et al. 1991; Nyman and Wallin 1992; Akiyama et al. 2001), suggesting a significant loss of content following polyploidization. Such reductions in genome size upon polyploidization have been well described in other plant species (Ozkan et al. 2003).

While quantitative descriptions continue to improve, recent studies have provided the first qualitative glimpse of genome composition. Gene density and total gene content was estimated by Pontaroli et al. (2009), extrapolated from Sanger sequencing 30 random 30–50 kb fosmid sequences using a GeneTrek approach (Bennetzen 2003). The study indicated somewhere between 24,600 and 35,500 genes in strawberry and approximately 4,500 truncated gene sequences. The genome contains representation of many transposable element classes, including *Copia* and *Gypsy* like LTR retrotransposons, non-LTR retrotransposons, MITES, CACTA-like transposons and *Mu*-like transposons. Gypsy-like elements comprise more than any other class (6.6%) in terms of total base-pair contribution to the genome. This study provides a basis for further comparisons to other genomes and studies of genome evolution.

Direct comparison of extant diploid and octoploid genotypes and phenotypes provides hints of polyploid origins, yet the precise constitution of the higher-ploid species remains to be resolved. As shown in Table 4.1-2, there are many diploid species comprised of many subspecies. Since the origin of the *Fragaria* genus is monophyletic (Rousseau-Gueutin et al. 2009), all of the modern octoploids share a common ancestor with the wild diploids. At some point in time a polyploidization event occurred and the diploid and polyploid parted their evolutionary journeys. Afterwards, crossing with other tetraploids or possibly led to the contemporary octoploid, accounting for cultivated strawberry's reticulate ancestry. Can we now use modern tools to dissect the events within *Fragaria* evolution, defining how, and possibly when polyploidization occurred?

The first evidence to resolve this mystery came from Ichijima's examination of meiotic pairings which presented some evidence of polyploid foundations noted earlier, indicating positive bivalent formation between *F. virginiana* (a wild octoploid) and *F. vesca bracteata* (a wild diploid). Many models have been presented since, suggesting allopolyploid, autopolyploid or mixed organization. Even those dating back to the 1940s recognize an organization based on multiple subgenomes contributing to the whole. Federova (1946) proposed a scheme of AAAABBCC. Senanayake and Bringhurst (1967) modified this formula as AAA'A'BBBB in order to take account of the evolutionary proximity of genomes A and C. Then, subsequent genetic evidence, notably the growing accumulation of disomic segregation data, prompted the proposal of the prevailing, fully differentiated AAA'A'BBB'B' model (Bringhurst 1990). Even to date, despite the plethora of molecular tools and phylogenic analyses of the ITS and chloroplast sequences (Potter et al. 2000), there is no firm and complete description of the relationship between the diploids and octoploid subgenome constituents. Recently, using two genomic genes and a phylogenetic approach, the genomic formula for the octoploid species (*F. virginiana* and *F. chiloensis*) may be hypothesized as follows: Y1Y1Y1Y1ZZZZ or Y1'Y1'Y1'Y1''ZZZZ, with the genomic formula of the tetraploid maternal donor as Y1Y1Y1Y1 or Y1'Y1'Y1'Y1'' (A or A' in the Bringhurst hypothesis for an ancestor of *F. vesca* or *F. mandshurica* as the donor) and the genomic composition of the other ancestral tetraploid candidate would be most presumably ZZZZ (B for *F. iinumae* or an ancestor of this species) (Rousseau-Gueutin et al. 2009). This genomic composition needs however to be further adjusted to each species, especially in the case of *F. iturupensis* for which both octoploid and decaploid numbers have been reported (Staudt 1973; Hummer et al. 2009).

The precise identity of these ancient donors is still a subject of inquiry and discussion, with many exciting lines of evidence circulating in conference abstracts and in collegial discourse. Several efforts have used sequencing of coding sequences, intron size, intergenic sequences and the ribosomal internal transcribed spacer to test this relationship. Abstracts from DeMeglio and Davis, presented at the Plant Speciation Symposium at St. Francis Xavier University, Antigonish, Nova Scotia, June 26–28, 2003, and as part of the Plant and Animal Genome Conference in San Diego, January, 2004, reported intron-length polymorphism of the *ALCOHOL DEHYDROGENASE* locus that diagnostically matched *F. iinumae*, *F. bucharica*, *F. mandshurica* and *F. vesca* to alleles present in the cultivated strawberry genome. A detailed study by Rousseau-Gueutin et al. (2009) analyzed the alleles of two low-copy nuclear genes (*GBSSI-2* and *DHAR*) in a comprehensive set of diploids, tetraploids, the hexaploid and wild octoploids. The results indicated that the genomes could be delineated into three subtypes. The first contains only *F. iinumae*, the next *F. vesca* and *F. mandshurica*, while *F. daltoniana*, *F. nilgerrensis*, *F. nipponica*, *F. nubicola*, *F. pentaphylla* and *F. yezoensis* are included in a third one. Within this latter group, *F. nilgerrensis* appeared as a well-differentiated evolutionary unit. In addition, in this study, the diploid *F. viridis* appeared to be linked to *F. vesca* and *F. mandshurica* while the diploid *F. bucharica* seemed to be a hybrid or issued from introgression between the two latter groups. The study relates these relationships to the octoploid construction, showing that an ancestor of both *F. iinumae* and either *F. vesca* or *F. mandshurica* were the likely subgenome donors, the latter possibly occurring via an *F. orientalis* tetraploid intermediate. An alternative approach employed the Gene Pair Haplotype (GPH), a concept that describes a unique character state of multiple markers (SNPs, indels or RFLPs) produced from an intergenic region (Davis et al. 2007). The hypothesis is that this rich combination of variable characters lends to a high resolution signature for a given genotype that possibly could be used to directly relate the diploids and discrete subgenomes of higher-ploids. Figure 4.2-2 shows sequence from intergenic sequence from a series of diploids and some octoploids, indicating these relationships and agreement with other subgenome assignments. The conserved fingerprints of diploid donors may be resolved in the context of the cultivated strawberry, providing evidence that these methods may unravel octoploid origins. All of the aforementioned studies do agree in that the diploids tend to resemble each other much more than they resemble the higher-ploids, so substantial sequence divergence has occurred since polyploidization, possibly accelerated by cultivation and selection. Currently this topic is being investigated with the power of high-throughput sequencing. Together these new tools will help dissect this ancient horticultural mystery.

4.2.4 Inheritance in Octoploid Strawberry

To even the casual student of biology the careful choreography of chromosome segregation is an awesome process. The process of meiosis maintains the facets of crossing over between homologous partners and precise compartmentalization of genetic materials. The complexity of the process increases unimaginably in the context of the octoploid. Not only do the number of genomes increase, they are typically then adjusted by diploidization, a process whereby gene redundancy is reduced via silencing, sequence elimination, rearrangement, retroelements and relaxing of imprinting (see reviews in Chen 2007). In the octoploid strawberry, according to most models, there is strict disomic inheritance—subgenomes

```
                                                                                    1
F. mandshurica  ATATGATGGGTAGATCGGGAGAGATAAAATTACCTGAATCTGAAGTGGGGGAAGTGAGTC
F. vesca        ATATGATGGGTAGATCGGGAGAGATAAAATTACCTGAATCTGAAGTGGGGGAAGTGAGTC
F. bucharica    ATATGATGGGTAGATCGGGAGAGATAAAATTACCTGAATCTGAAGTGGGGGAAGTGAGTC
F. ×ananassa    ATATGATGGGTAGATCGGGAGAGATAAAATTACCTGAATCTGAAGTGGGGGAAGTGAGTC
F. iinumae      ATATGATGGGTAGATCGGGAGAGATAAAATTACCAGAATCTGAAGTGGGGGAAGTGAATC
                                                       2                    3
F. mandshurica  AGTGAAGGACTGAGTTGGTGGAGTCTTGGGAGATCTGAGATATGAGCTCTAAAGCCGGCG
F. vesca        AGTGAAGGACTGAGTTGGTGGAGTCTTGGGAGATCTGAGATATGAGCTCTAAAGCCGGCG
F. bucharica    AGTGAAGGACTGAGTTGGTGGAGTCTTGGGAGATCTGAGATATGAGCTCTAAAGCCGGCG
F. ×ananassa    AGTGAAGGACTGAGTTGCTGGAGTCGTGGGAGATC-------TGAGC-------CCGGCG
F. iinumae      AGTGAAGGACTGAGTTGCTGGAGTCTTGGGAGATC-------TGAGCTCTAAAGCCGGCG
                    1
F. mandshurica  AAAGGATGCGCGGCGCAGGATAGGAGGGAACAGGGTGCGTAGGATAACCCAATCAATGAA
F. vesca        AAAGGATGCGCGGCGCAGGATAGGAGGGAAAAGGGTGCGTAGGATAACCCAATCAATGAA
F. bucharica    AAAGGATGCGCGGCGCAGGATAGGAGGGAAAAGGGTGCGTAGGATAACCCAATCAATGAA
F. ×ananassa    AAAGGATGCGCGGCGCAGGATAGGAGGGAAAAGGGTGCGTAGGATAACCCAATCAATGAA
F. iinumae      AAAGGATGCGCGGCGCAGGATAGGAGGGAAAAGGGTGCGTAGGATAACCCAATCAATGAA
                1                                                          1
F. mandshurica  CCAAATGAGAATACGCTAGTGATTTTGATTATGAATTCTATAAATTCTATAAAAA-TTTA
F. vesca        CCAGATGAGAATACGCTAGTGATTTTGATTATGAATTCTATAAATTCTATAAAAA-TTTA
F. bucharica    CCAAATGAGAATACGCTAGTGATTTTGATTATGAATTCTATAAATTCTACAAAAAATTTA
F. ×ananassa    CCAAATGAGAATACGCTAGTGATTTTGATTATGAATTCTATAAATTCTACAAAAA-TTTA
F. iinumae      CCAAATGAGAATACGCTAGTGATTTTGATTATGAATTCTATAAATTCTACAAAAA-TTTA

F. mandshurica  TTTCATTTCTTAATTCTTACTCTGTTTCGGTGTTGGCCAGATTTGACTCTTCTGTGCTTC
F. vesca        TTTCATTTCTTAATTCTTACTCTGTTTCGGTGTTGGCCAGATTTGACTCTTCTGTGCTTC
F. bucharica    TTTCATTTCTTAATTCTTACTCTGTTTCGGTGTTGGCCAGATTTGACTCTTCTGTGCTTC
F. ×ananassa    TTTCATTTCTTAATTCTTACTCTGTTTCGGTGTTGGCCAGATTTGACTCTTCTGTGCTTC
F. iinumae      TTTCATTTCTTAATTCTTACTCTGTTTCGGTGTTGGCCAGATTTGACTCTTCTGTGCTTC
                  2      1    3        3                        3
F. mandshurica  AGT------TTTGACCATTTATTTTATATCCTCAGGAAGGGTTCAAGCGCGGCCTGCCA
F. vesca        AGT------TTTGACCATTTACTTTTATAACCTCAGGAAGGGTTCAAGCGCGGCCTGCCA
F. bucharica    AGT------TTTGACCATTTACATTTATAACCCCGGGAAGGGTTCAAGCGCGGCCTGCCA
F. ×ananassa    AGTCATGACTTTGACCATTTACTTT-----------AGGGTTCAAGCGCAGCCTGCCA
F. iinumae      AGTCATGACTTTGACCATTTACTTTTATAACCCCAGGAAGGGTTCAAGCGCGGCCTGCCA
                  2           1            3       3          3     3
F. mandshurica  CGTGGTGAATTC-----------AAAAGAGTCTGGAAGCAAAGCCTTGACCTCGTGGAA
F. vesca        CGTGGTGAATTC-----------AAAAGAGTCTGGAAGCAAAGCCTTGACCTCGTGGAA
F. bucharica    CGTGGTGAATTC-----------AAAAGAGTCTGGAAGCAAAGCCTTGACCTCGTGGAA
F. ×ananassa    CGTGGTGAATTCTGGTTCGTCCGGAAAAGAGCCTGGAAGAAAAGCCTTGACCTCGTGGAT
F. iinumae      CGTGGTGAATTCTGGTTCGTCCGGAAAAGAGTCTGGAAGCAAAGCCTTGACCTCGTGGAA
                  3           1 1                       1
F. mandshurica  TTCGTCTCTCCCCTCCCGGT-ACAGTAACTTTATCGTTTTACCGCTAGTATGTCTCTGTC
F. vesca        TTCGTCTCTCCCCTCCCGGTAACAGTAACTTTATCG----------------------
F. bucharica    TTCGTCTCTCCCCTCCCGGTAACAGTAACTTTATCGTTTTACCGCTAGTATGTCTCTGTC
F. ×ananassa    TTCGTTTCTCCCCTCCCGGTAACAGTAACTTTATCGTTTTACCGCTAGTATGTCTCTGTC
F. iinumae      TTCGTCTCTCCCCTCCCGGGTAACAGTAACTTTATCGTTTTACCGCTAGCATGTCTCTGTC
```

Figure 4.2-2 CLUSTALW alignments of four diploid strawberry accessions and one allele from "Strawberry Festival", an octoploid. The region is part of an intergenic region amplified by PCR. Several conserved allele states are shown in gray. 1. SNP identity between octoploid and a subset of diploids. 2. An indel shared between cultivated strawberry and *F. iinumae*. 3. An indel shared between all diploids, not common to this octoploid allele. Such analyses when done more completely allow associations between octoploid alleles and diploid wild species, inferring evolutionary relationships.

segregating only within themselves. Here four independent sets of chromosomes somehow successfully find their cognate and pass genetic information accordingly—or so it is thought. A number of lines of data supported contrasting interpretations regarding disomic or polysomic inheritance patterns in octoploid strawberry, but today a firm conclusion of major or exclusive disomic inheritance may be drawn.

In agreement with genome structure models, several molecular and biochemical markers suggested disomic inheritance; that the subgenomes of the octoploid strawberry acted as individual diploids. Analysis of isozyme variants (Arulsekar and Bringhurst 1981b) supported disomic segregation, as did analysis of microsatellite marker segregation in wild populations of *F. virginiana* as well as controlled crosses (Ashley et al. 2003). These results were confirmed by Kunihisa et al. (2005) in their analysis of a single locus CAPS marker that segregated as a disomic locus in the octoploid background.

In a first study using an octoploid segregating population (Lerceteau Köhler et al. 2003) with AFLP markers, analyses of the markers in coupling and repulsion phase suggested mixed segregation, not entirely disomic as proposed by others. When this relationship was re-examined by the same group by analyzing the inheritance of simple sequence repeat (SSR) markers, despite the possible existence of residual levels of polysomic segregation suggested by the observation of large linkage groups in coupling phase only, the prevalence of linkage groups in coupling/repulsion phase

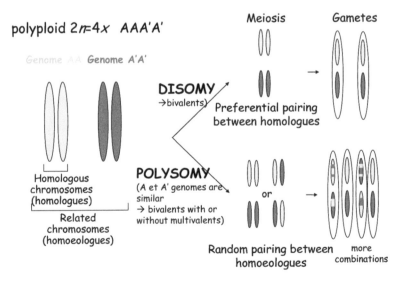

Figure 4.2-3 An example of chromosome pairing at metiosis in a tetraploid (2n = 4x) AAA′A′ under disomic and polysomic behavior.

Color image of this figure appears in the color plate section at the end of the book.

clearly demonstrates that the meiotic behavior is mainly disomic in the cultivated strawberry (Rousseau-Gueutin et al. 2008). The difference was easily reconcilable, as the latter study was based on a large number of markers, included a robust codominant microsatellite, implemented larger populations, and produced an accounting of the complete expected set of octoploid linkage groups. The larger number of markers spanning the entire genome renders this latter study a conclusive one, strongly supporting the hypothesis of disomic segregation in cultivated strawberry.

4.2.5 Molecular Markers—Fingerprints and Map Making

Tangible products of strawberry genetics and genomics are best evident when they translate to the grower's field. To make this step it is important to define where genes (or specific alleles of a gene) reside in a genome and/or develop high-resolution and reliable means to track them through generations. Molecular markers rise to this occasion. Molecular markers are simply reproducible products of an assay intended to identify variability between individuals. In contemporary parlance these are nucleic acid based, although enzyme variants could be (and have been) associated with a given trait. Such tools are extremely valuable to breeders, especially because production is dependent upon a synthesized output of quantitative traits from several subgenomes. A durable biomarker that segregates faithfully with a trait of interest is extremely valuable. The widespread implementation of molecular markers has not been practiced in strawberry, at least in public breeding programs. To the contrary, private breeding programs have realized the power of molecular technologies in significantly limiting the number of plant lines to be screened each year. While there are some excellent examples of success stories in defining marker-trait associations (reviewed later in this chapter) the main use has been to use them as tools to describe the strawberry genome.

The process of marker development has been most frequently implemented as the foundation of cultivar fingerprinting and linkage mapping. Marker development in strawberry has been well described in various reviews of literature (Hokanson 2001; Hokanson and Maas 2001; Davis et al. 2006; Folta and Davis 2006), so treatment herein will only present a distillation of major concepts and highlights.

Polymorphic isozyme markers were used for study of inheritance, for differentiation of strawberry cultivars and somaclonal variation. The diploidized nature of the octoploid strawberry was shown by Arulsekar et al. (1981) using phosphoglucoisomerase (PGI) and leucine amino peptidase (LAP) and further studies provided a genetic model for PGI markers in *F. vesca* (Arulsekar and Bringhurst 1981). Distinction between genotypes of the cultivated octoploid *Fragaria* has been performed using mainly the three following isoenzymes PGI, LAP, and phosphoglucomutase (PGM) (Bell and Simpson 1994), even recently in addition to DNA molecular markers. Similar studies using isoenzymes in addition to DNA molecular markers have also been conducted on the Chilean strawberries, *F. chiloensis* Gambardella et al. 2005). In addition, variations of somaclones from the strawberry cultivar have been also investigated through biochemical variation using isoenzymes (Flores et al. 2000).

An array of molecular techniques has been used in strawberry marker development, mostly based on PCR-driven strategies. Early studies utilized a series of molecular marker types to fingerprint germplasm and diagram genetic diversity analysis (Graham et al. 1996; Degani et al. 1998, 2001; Tyrka et al. 2002; Arnau et al. 2003; Debnath et al. 2008; Govan et al. 2008). Although the reports are plentiful, the use of PCR-based approaches such as randomly amplified polymorphic DNA (RAPD) or AFLPs present challenges in precise processes like fingerprinting. These methods are arbitrary, keying off of sequences in the genome that alter primer binding or amplicon size. These events can be highly influenced by DNA quality, ions, temperature, and a host of other variables that may not be easy to control from laboratory to laboratory, limiting their application. While random PCR-based marker methods are considered generally unreliable, full disclosure dictates noting the clear successes where these methods have been applied. RAPDs have been used in cultivar identification (Korbin et al. 2002), genetic diversity (Kuras et al. 2004) and underlie the foundation of the *Fragaria vesca* linkage map (Davis and Yu 1997). AFLP-based markers were used to devise a highly-complex linkage map in *F.* × *ananassa* (Lerceteau-Köhler et al. 2003). It has been noted that the more random marker strategies are most useful when complementing other methods (Korbin et al. 2002).

The use of SSRs (or microsatellites) has proven to be a much more effective and transferable means to answer questions in strawberries. SSRs represent the variable expansion of polynucleotide repeats, small islands of redundant sequences that rapidly expand and contract with time. Because they originate from primer sequences with known binding sites and known amplicon sizes, they can be reliably (and inexpensively) amplified and analyzed. Variability is rich across strawberry accessions (as an example see Sánchez-Sevilla et al. 2009 for the cultivated strawberry genetic diversity). Here the octoploid genome is for once an asset, as the allelic diversity in a polyploid produces complex band patterns that may serve as important forensic markers for linkage assignments or for plant authentication (Fig. 4.2-4).

In *Fragaria*, microsatellites have been developed from species of different ploidy levels: from the cultivated octoploid strawberry, *F.* × *ananassa* (Davis et al. 2006; Rousseau-Gueutin et al. 2008), from the octoploid *F. virginiana* (Ashley et al. 2003), from three diploid species, *F. vesca* (James et al. 2003;

Cipriani and Testolin 2004; Hadonou et al. 2004), and *F. viridis* (Sargent et al. 2003). The markers derived from expressed sequence tags (EST) based SSRs seem to be more transferable than those designed from SSR-enriched genomic libraries (Bouck and Vision 2007). Development of such EST-SSR markers was initiated in diploid and in octoploid *Fragaria* species (Folta et al. 2005; Bassil et al. 2006a, b; Gil-Ariza et al. 2006; Sargent et al. 2006b).

The transferability of genomic SSRs was reported to be high within the genus *Fragaria*, with 70 to 100% being transferable (Ashley et al. 2003; Bassil et al. 2006a; Davis et al. 2006; Rousseau-Gueutin et al. 2008). A substantial majority of primers derived from the cultivated octoploid successfully amplified products in diploid species, with *F. vesca* the most compatible (98.4% of primer pairs) and *F. viridis* the least (73.4%) (Davis et al. 2006). Ultimately only 20–30% could be transferred to other genera in the Rosoideae, such as *Rosa* (Zorrilla-Fontanesi et al. 2010; Rousseau-Gueutin et al. 2010) and *Rubus* (Stafne et al. 2005). Low cross-genera transference in the Rosoideae was also observed using EST-SSR markers due for example to the absence of the microsatellite motifs (Rousseau-Gueutin et al. 2010). All these microsatellite markers are useful for comparative mapping within a genus and sometimes between genera.

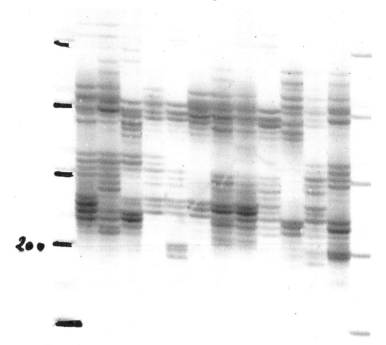

Figure 4.2-4 Example of acrylamide gel patterns produced by amplification of the SSR UDF005 (Cipriani and Testolin 2004) on 12 genotypes belonging to the cultivated strawberry, *F. × ananassa*.

With the large expansion of DNA, particularly EST, sequence databases, the research of gene specific markers has been developed. Polymorphisms of the genes were visualized by different methods such as indels in introns (Deng and Davis 2001), single strand conformation polymorphism (SSCP) (B. Denoyes Rothan, unpubl results) or cleaved amplified polymorphic sequences (CAPS) (Kunihisa et al. 2003). With the low cost of sequencing, polymorphism on candidate genes is now based on single nucleotide polymorphism (SNP). These SNPs have been recently developed for linkage map (Sargent et al. 2004, 2006b). Currently attemps are being made to develop SNP resources for octoploid *Fragaria*. This approach claims to develop SNPs that amplify only one of the four sub-genomes or to detect the level of dose, from single-dose to multi-dose, of a given SNP marker using (for example) HRM techniques.

4.2.6 Development of Linkage Maps

Linkage maps in *Fragaria* have been developed for strawberry at different levels of ploidy. The first linkage map reported in strawberry was constructed using RAPD markers in an F_2 population generated from a cross of *F. vesca* ssp. *vesca* "Baron Solemacher", a non-runnering European cultivar, and *F. vesca* ssp. *americana* W6, a wild runnering accession collected in New Hampshire (Davis and Yu 1997). The map covered 445 cM, divided among the anticipated seven linkage groups, and was comprised of 75 RAPDs, the two previously described isozymes, the runnering locus, and an STS marker based on the alcohol dehydrogenase gene.

Deng and Davis (2001) constructed a minimal diploid map in order to assign six genes of the anthocyanin pathway to the appropriate linkage groups. Using RAPD markers mapped in the previous study by Davis and Yu (1997) they placed the genes on five of the previously defined linkage groups, along with the previously mapped color locus, *c*, using two populations: *F. vesca* ssp. *vesca* "Yellow Wonder" × FRA520 (thought at the time to be *F. nubicola* but now believed to be *F. bucharica* (G. Staudt, pers comm) and *F. vesca* ssp. *bracteata* DN1C (a wild accession from Northern California) × "Yellow Wonder".

A very productive F_2 population has contributed greatly to the current reference map which came from a cross of *F. vesca* '815' and *F. bucharica* '601', as described by Sargent et al. (2006a, b). As initially the published map consisted of 68 microsatellites, one sequence-characterized amplified region (SCAR), six gene-specific markers, and three morphological traits, including the previously uncharacterized *pale-green leaf* (*pg*) marker. While covering a similar distance (448 cM) and number of markers (78), as the previous diploid map by Davis and Yu (1997), the reliance on more transferable types of markers provides a potentially more useful framework for mapping in other populations, including the octoploid.

Sargent et al. published a further extension of this map in 2006, adding 109 additional SSR markers, including 45 newly developed ones, and again in 2007 with 24 new gene specific markers (Sargent et al. 2007). A second mapping population, a BC_1 of *F. vesca* FRA815 × (*F. vesca* FRA815 × *F. viridis* FRA903), was created using a selection of 33 markers from the FV×FN map. This map demonstrated that marker order was preserved even in the more distantly related *F. viridis* (Nier et al. 2006), giving hope that established diploid maps might readily be adapted to the octoploid.

Segregating populations have been developed at the octoploid level, mainly in the cultivated octoploid species, *F. × ananassa*. At the beginning of this development, no clear information on the meiotic behavior of the cultivated *Fragaria* genome was firmly understood, despite hypotheses on disomy for some markers (Ashley et al. 2003; Kunihisa et al. 2005). In this context, pseudo-test cross populations were obtained in order to construct the map according to the multistep process developed by Wu et al. (1992), and then by Ripol et al. (1999) and Qu and Hancock (2001). Segregation of each allele is analyzed and only single dose alleles are used for construction of the linkage maps. Linkages in coupling and repulsion phases are subsequently identified. This approach has been developed in the auto-allopolyploid sugarcane (Da Silva et al. 1995; Grivet et al. 1996; Aitken et al. 2005). Disomy could be distinguished from polysomy by comparison of the number of loci or linkage groups linked in coupling versus repulsion phase (Sorrells 1992; Wu et al. 1992; Da Silva et al. 1993) or by analysis of the ratio of single to multiple dose markers (Da Silva et al. 1993).

The first genetic map in polyploid strawberry was obtained using AFLP markers, constructed using a two-way pseudo-test cross of "Capitola" × CF1116, and generated female and male maps with linkage groups at 1,604 cM and 1,496 cM in length, respectively (Lerceteau-Köhler et al. 2003). The linkage associations identified were later expanded using SSR markers (Rousseau-Gueutin et al. 2008). More recently, two other linkage maps were obtained using AFLP markers (Weebadde et al. 2008) or SSR markers (Sargent et al. 2009) with the goal of mapping quantitative trait loci (QTL) loci associated with day neutrality (Fig. 4.2-5). Comparative mapping beween diploid and octoploid linkage maps revealed high levels of macrosynteny and colinearity, which suggested that polyploidization did not trigger any major chromosomal rearrangement within *F. × ananassa* (Rousseau-Gueutin et al. 2008, Sargent et al. 2009) (Fig. 4.2-5).

4.2.7 Germplasm Identification

This use of SSRs in fingerprinting is shown brilliantly in a recent study by Govan et al. (2008), where they tested a suite of SSR primers effective in linkage mapping for genotyping with microsatellites. Linkage analysis identified primer sets that were most robust in the reaction and offered the highest likelihood of successful amplification over a range of cultivars. The study tested over 60 octoploid genotypes, and narrowed the set of primers to 10 core primer pairs. These 10 provided highly variable, yet reproducible, amplicon families that could be used to distinguish one octoploid strawberry line from another, even related lines. This same primer set has been adopted to fingerprint the University of Florida strawberry germplasm and will likely be useful in standardizing a set of fingerprints in the species.

4.2.8 Mapping and Tagging of Major Genes

Development of molecular markers linked to major genes in cultivated strawberry has lagged behind that of other similarly important crops for perhaps two reasons. First, the complex polyploid nature of the crop is such that very few traits are truly monogenic, and second, the small space requirements and quick life cycle mean that for many traits it is often cheaper

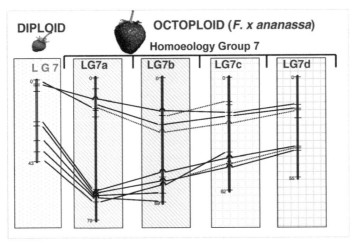

Cytological formula of *F. x ananassa* could be: [AAA'A'BBBB], [Y1'Y1'Y1"Y1"ZZZZ]
or [AAA'A'BBB'B], [Y1Y1Y1Y1ZZZZ]
- where A or Y is *F. vesca*, *F. mandshurica* or an ancestor of them
- and B, B' or Z represent *F. iinumae* or its ancestor

Figure 4.2-5 High levels of conserved macrosynteny and colinearity within homo(eo)logous linkage group 7 of *F. × ananassa* and between this octoploid homoeologous group and its corresponding diploid linkage group 7 from the reference diploid linkage map. Two of the four sub-genomes, which would belong to *F. vesca* or *F. mandshurica* or their ancestor, are indicated using slanting lines while the two other sub-genomes, which can be attributed to *F. iinumae* or its ancestor, are indicated using crossed lines.

Color image of this figure appears in the color plate section at the end of the book.

and easier to simply grow out and evaluate seedlings than it is to evaluate them molecularly. Still, a number of markers have been developed, and at least one breeding program, that of Driscoll Strawberry Associates, has been actively utilizing proprietary markers in their breeding program.

Several such markers have been developed for disease resistance genes. Haymes et al. (1997) developed RAPD markers linked to the *Rpf1* gene, which confers dominant resistance to several races of *Phytophthora fragariae*. These markers were later converted into SCAR markers more readily useful to breeding programs (Haymes et al. 2000; Rugienius et al. 2006). Further work by Haymes et al. also revealed RAPD markers linked to three other *P. fragariae* resistance genes, *Rpf2, Rpf3,* and *Rpf6,* (Haymes 2000).

Lerceteau-Köhler et al. (2003) conducted a QTL analysis of fruit traits over two years, utilizing the octoploid "Capitola" × CF1116 mapping population previously described above. A total of 22 significant QTLs the first year, and 17 the second, were detected, but only two were significant in both years. The percentages of phenotypic variance explained by each QTL were relatively small, ranging from 6.5% to 16.0%.

Markers have also been associated with resistance to one of strawberry's major diseases. SCAR markers linked to the *Rca2* gene, a source of resistance to *Colletotrichum acutatum*, were developed from RAPDs. Two SCARs were developed, at 2.8 and 0.8 cM from the major gene. These were tested across a wide range of cultivars and found to be fairly reliable, although crown resistance, which was the basis of the screen, and fruit resistance were not always consistently linked (Denoyes-Rothan et al. 2005; Lerceteau-Köhler et al. 2005).

Outside of gene resistance traits, the everbearing trait is the only major gene associated with molecular markers. Sugimoto et al. (2005) developed several RAPD markers showing moderate linkage to the everbearing habit in a cross of "Ever Berry" × "Toyonoka". However, although in this cross and in previous work in this background (Monma et al. 1990; Igrashi et al. 1994) this trait has appeared to segregate as a single dominant gene. More recent works by Shaw and Famula (Igrashi et al. 1994; Shaw 2003; Shaw and Famula 2005) and Weebadde et al. (2008) suggest a more complex basis for this trait. The latter report mapped several QTLs associated with this trait in a cross of Honeoye × Tribute. Progeny of this cross were grown in five independent locations across the US and scored for blooming under the long days of summer. Eight QTLs associated with everbearing habit were observed, and one of these explained until 36% of phenotypic variation in three (MI, MN, and MD) of the five locations studied. Similarly, a locus associated with the trait in California was absent in the other sites. The differences between the single-locus detected by Sugimoto et al. (2005) and other works suggesting polygenic control can possibly be ascribed to day-neutrality source. The "Ever Berry" cultivar appears to derive its everbearing habit from "Ooishi-Shikinari" an older everbearing Japanese cultivar which likely derives the trait from the older European or American everbearing source, rather than the *F. virginiana* ssp. *glauca* source used in modern day-neutral breeding and in the Shaw and Famula (2005) and Weebadde et al. (2008) studies.

Other attempts at developing markers for flowering were attempted using the diploid populations. Using the FV × FN population, Sargent et al. (2006a) examined six phenotypic traits and detected nine significant (LOD > 2.0) QTLs, explaining between 10.6% and 30.3% of the phenotypic variance for the associated traits. When placed on the linkage map for this population, four of these QTLs clustered in a single area of LG VI, near the *SFL* locus. None of these traits (leaflet length, flower diameter, receptacle diameter, and stolon internode length) would seem to necessarily associate with seasonal flowering, and, surprisingly, the one trait measured that might be assumed to be affected by the trait, date of first flowering, did not show a significant association with this locus.

4.2.9 The Strawberry Genome

Historically speaking, promises of strawberry genome sequencing have been punctuated with disappointment. Researchers agreed that the genome was small, a hint larger than *Arabidopsis thaliana*. As far as genome size goes this is just the next step. Unfortunately, federal agencies and the community at large were not interested in financing the sequencing of strawberry, for a multitude of sound reasons. There was no physical map, the linkage maps were excellent but did not offer sufficient anchorage for sequence assemblies. Most importantly, there were no genomics resources such as bacterial artificial chromosome (BAC) libraries to actually sequence. One would be hard pressed to get a consensus of researchers to agree on the most suitable accession to sequence. While genome sequence was critical in sorting out the discrepancies in strawberry evolution and development of subgenome-specific markers, such activities would have to be set aside until technology could catch up. The topic is covered in some depth in the following chapter. In all estimations the final draft sequence of the *F. vesca* Hawaii-4 accession should be published concurrently with this volume.

Aside from the reference sequence of a diploid (the accession Hawaii-4), it will be difficult to learn much about strawberry genome structure from massive short read sequencing of octoploids. The genomes are similar enough to preclude simple assignment of any given short sequence to one subgenome or the other, and the possibility of multiple alleles at a given locus and the abundance of transposable elements further clouds this possibility. The only real utility will be to describe homozygosity, as individual potential subgenome donors may be sequenced and compared to the findings in the octoploids. Longer-reads, single molecule sequencing, and methods not yet dreamed of may assist in this process. For now, the process only promises to allow researchers to survey the diversity of potential alleles in candidate genes within these genomes and perhaps resort to older technologies to tackle these individual questions.

4.2.10 Conclusions and Beginnings

The publication of a complete strawberry genome sequence is not the beginning of the end of strawberry research. Instead, it is the end of the beginning. With a complete parts list it is now time to use the suite of useful transgenic systems to rapidly elucidate the function of each one of these genes. T-DNA insertion lines, overexpression and RNAi and activation tagging approaches are currently being implemented in a number of laboratories all over the globe. With its rapid cycling, ease of transformation, and the most modern tools (such as rapid recombination-based cloning methods), it will be possible to quickly ascribe function to strawberry genes. The tsunami of functional data is on the horizon, initiated with a foundation of understanding the genetics and general genomics of strawberry.

How far have we come in unraveling the genetics and genome structure of this novel organism? One hundred years before Mendel, Duchesne crossed strawberries and observed the transmission of traits in the resulting progeny. His keen artistic and scientific eye probably witnessed specific patterns of inheritance, but the observations were likely not in accord with simple mathematical ratios. In this way can we even fathom where molecular-genetic sciences would be if Duchesne had emphasized study of the diploids rather than trials in the octoploids?

Even in the year 2000 the best genetic linkage relationships defined only a few physical associations and strawberry functional genomics was confined to a few studies of a few genes and a GenBank holding only a dozen deposits. One decade later our understanding of *Fragaria* genetics and the *Fragaria* genome are just reaching a critical mass that will speed discovery into the next 10 years. These discoveries eventually will place a thicker textbook on the library shelf and a better strawberry in the grocery store.

References

Aitken KS, Jackson PA, McIntyre CL (2005) A combination of AFLP and SSR markers provides extensive map coverage and identification of homo(eo) logous linkage groups in a sugarcane cultivar. Theor Appl Genet 110: 789–801.

Akiyama Y, Yamamoto Y, Ohmido N, Oshima M, Fukui K (2001) Estimation of the nuclear DNA content of strawberries (*Fragaria* spp.) compared with *Arabidopsis thaliana* by using dual-stem flow cytometry. Cytologia 66: 431–436.

Arnau G, Lallemand J, Bourgoin M (2003) Fast and reliable strawberry cultivar identification using inter simple sequence repeat (ISSR) amplification. Euphytica 129: 69–79.

Arulsekar S, Bringhurst RS (1981) Genetic model for the enzyme marker PGL in diploid *CALIFORNIA FRAGARIA-VESCA* L—Its variability and use in elucidating the mating system. J Hered 72: 117–120.

Arulsekar S, Bringhurst R, Voth V (1981) Inheritance of PGI and LAP isozymes in octoploid cultivated strawberries. J Am Soc Hort Sci 106: 679–683.

Ashley MV, Wilk JA, Styan SM, Craft KJ, Jones KL, Feldheim KA, Lewers KS, Ashman TL (2003) High variability and disomic segregation of microsatellites in the octoploid *Fragaria virginiana* Mill. (Rosaceae). Theor Appl Genet 107: 1201–1207.

Bassil NV, Gunn M, Folta K, Lewers K (2006a) Microsatellite markers for Fragaria from 'Strawberry Festival' expressed sequence tags. Mol Ecol Notes 6: 473–476.

Bassil NV, Njuguna W, Slovin JP (2006b) EST-SSR markers from *Fragaria vesca* L. cv. yellow wonder. Mol Ecol Notes 6: 806–809.

Bell JA, Simpson DW (1994) The use of isoenzyme polymorphisms as an aid for cultivar identification in strawberry. Euphytica 77: 113–117.

Bennett MD, Leitch IJ, Price HJ, Johnston JS (2003) Comparisons with *Caenorhabditis* (~100 Mb) and *Drosophila* (~ 175 Mb) using flow cytometry show genome size in Arabidopsis to be ~ 157 Mb and thus approximately 25% larger than the Arabidopsis genome initiative estimate of ~ 125 Mb. Ann Bot (Lond) 91: 547–557.

Bennetzen J (2003) Higher throughput comparative genomics of wheat and related cereals. Proc 10th Intl Wheat Genet Symp 1: 215–220.

Bors B, Sullivan JA (1998) Interspecific crossability of nine diploid Fragaria species. HortScience 32: 439.

Bouck A, Vision T (2007) The molecular ecologist's guide to expressed sequence tags. Mol Ecol 16: 907–924.

Bowers JE, Chapman BA, Rong J, Paterson AH (2003) Unravelling angiosperm genome evolution by phylogenetic analysis of chromosomal duplication events. Nature 422: 433–438.

Bringhurst RS (1990) Cytogenetics and Evolution in American Fragaria. HortScience 25: 879–881.

Bringhurst R, Senanayake Y (1966) Evolutionary significance of natural *Fragaria chiloensis* x *F. vesca* hybrids resulting from unreduced gametes. Am J Bot 53: 1000-&.

Cipriani G, Testolin R (2004) Isolation and characterization of microsatellite loci in Fragaria. Mol Ecol Notes 4: 366–368.

Comai L (2005) The advantages and disadvantages of being polyploid. Nat Rev Genet 6: 836–846.

Cui L, Wall PK, Leebens-Mack JH, Lindsay BG, Soltis DE, Doyle JJ, Soltis PS, Carlson JE, Arumuganathan K, Barakat A, Albert VA, Ma H, dePamphilis CW (2006) Widespread genome duplications throughout the history of flowering plants. Genome Res 16: 738–749.

Darrow G (1966) The Strawberry: History Breeding and Physiology. Holt, Rinehart and Winston, New York, USA.

Dasilva JAG, Sorrells ME, Burnquist WL, Tanksley SD (1993) RFLP linkage map and genome analysis of *Saccharum spontaneum*. Genome 36: 782–791.

Dasilva J, Honeycutt RJ, Burnquist W, Aljanabi SM, Sorrells ME, Tanksley SD, Sobral BWS (1995) *Saccharum spontaneum* L Ses-208 genetic linkage map combining RFLP-based and PCR-based markers. Mol Breed 1: 165–179.

Davis T, DiMeglio L, Yang R, Styan S, KS L (2006) Assessment of SSR marker transfer from the cultivated strawberry to diploid strawberry species: functionality, linkage group assignment, and use in diversity analysis. J Am Soc Hort Sci Volume 131: 506–512.

Davis T, Folta KM, Shields M, Zhang Q (2007) Gene pair markers: An innovative tool for comparative linkage mapping. Proc N Am Strawerry Symp 7: 105–107.

Davis TM, Yu H (1997) A linkage map of the diploid strawberry, *Fragaria vesca*. J Hered 88: 215–221.

Debnath SC, Khanizadeh S, Jamieson AR, Kempler C (2008) Inter Simple Sequence Repeat (ISSR) markers to assess genetic diversity and relatedness within strawberry genotypes. Can J Plant Sci 88: 313–322.

Degani C, Rowland LJ, Levi A, Hortynski JA, Galletta GJ (1998) DNA fingerprinting of strawberry (Fragaria x ananassa) cultivars using randomly amplified polymorphic DNA (RAPD) markers. Euphytica 102: 247–253.

Degani C, Rowland LJ, Saunders JA, Hokanson SC, Ogden EL, Golan-Goldhirsh A, Galletta GJ (2001) A comparison of genetic relationship measures in strawberry (*Fragaria x ananassa* Duch.) based on AFLPs, RAPDs, and pedigree data. Euphytica 117: 1–12.

Deng C, Davis TM (2001) Molecular identification of the yellow fruit color (c) locus in diploid strawberry: a candidate gene approach. Theor Appl Genet 103(2-3): 316–322.

Denoyes-Rothan B, Guerin G, Lerceteau-Köhler E, Risser G (2005) Inheritance of resistance to *Colletotrichum acutatum* in *Fragaria x ananassa*. Phytopathology 95: 405–412.

Dickinson TA, Lo E, Talent N (2007) Polyploidy, reproductive biology, and Rosaceae: understanding evolution and making classifications. Plant Syst Evol 266: 59–78.

Fedorova N (1946) Crossability and phylogenetic relations in the main European species of *Fragaria*. Comp Rend Acad Sci USSR 53: 545–547.

Flores R, Rocha BG, Peters JA, Augustin E, Fortes RLF (2000) Isozymic analyses of somaclones from strawberry (Fragaria x ananassa Duch.) cv. Vila Nova Ciência Rural (Brazil) 30: 993–997.

Folta K, Staton M, Stewart P, Jung S, Bies D, Jesudurai C, Main D (2005) Expressed sequence tags (ESTs) and simple sequence repeat (SSR) markers from octoploid strawberry (*Fragaria x ananassa*). BMC Plant Biol 5: 12.

Folta KM, Davis TM (2006) Strawberry genes and genomics. CritRev Plant Sci 25: 399–415.

Gambardella M, Cadavid A, Diaz V, Pertuze R (2005) Molecular and morphological characterization of wild and cultivated native Fragaria in southern Chile. HortScience 40: 1640–1641.

Gil-Ariza DJ, Amaya I, Botella MA, Blanco JM, Caballero JL, Lopez-Aranda JM, Valpuesta V, Sanchez-Sevilla JF (2006) EST-derived polymorphic microsatellites from cultivated strawberry (*Fragaria x ananassa*) are useful for diversity studies and varietal identification among Fragaria species. Mol Ecol Notes 6: 1195–1197.

Gil-Ariza DJ, Amaya I, Lopez-Aranda JM, Sanchez-Sevilla, JF, Botella, MA, Valpuesta, V (2009) Impact of plant breeding on the genetic diversity of cultivated strawberry as revealed by EST-derived SSR markers. J Am Soc Hort Sci 134: 337–347.

Govan C, Simpson D, Johnson A, Tobutt KR, Sargent DJ (2008) A reliable multiplexed microsatellite set for genotyping *Fragaria* and its use in a survey of 60 *F. x ananassa* cultivars. Mol Breed 22: 649–661.

Graham J, McNicol RJ, McNicol JW (1996) A comparison of methods for the estimation of genetic diversity in strawberry cultivars. Theor Appl Genet 93: 402–406.

Grivet L, DHont A, Roques D, Feldmann P, Lanaud C, Glaszmann JC (1996) RFLP mapping in cultivated sugarcane (*Saccharum* spp.): Genome organization in a highly polyploid and aneuploid interspecific hybrid. Genetics 142: 987–1000.

Hadonou AM, Sargent DJ, Wilson F, James CM, Simpson DW (2004) Development of microsatellite markers in *Fragaria*, their use in genetic diversity analysis, and their potential for genetic linkage mapping. Genome 47: 429–438.

Hancock JF (1999) Strawberries. CABI Publ, New York, NY, USA.

Harrison RE, Luby JJ, Furnier GR (1997) Chloroplast DNA restriction fragment variation among strawberry (*Fragaria* spp.) taxa. J Am Soc Hort Sci 122: 63–68.

Haymes KM, Henken B, Davis TM, vandeWeg WE (1997) Identification of RAPD markers linked to a *Phytophthora fragariae* resistance gene (*Rpf1*) in the cultivated strawberry. Theoretical and Applied Genetics 94(8): 1097–1101.

Haymes KM, Van de Weg WE, Arens P, Maas JL, Vosman B, Den Nijs APM (2000) Development of SCAR markers linked to a *Phytophthora fragariae* resistance gene and their assessment in European and North American strawberry genotypes. J Am Soc Hort Sci 125(3): 330–339.

Hodgson J (2007) Evolution of sequencing technology. Science 316: 846a check.

Hokanson SC (2001) SNiPs, chips, BACs, and YACs: Are small fruits part of the party mix? HortScience 36: 859–871.

Hokanson SC, Maas JL (2001) Strawberry biotechnology. In: J Janick (ed) Plant Breeding Reviews. John Wiley and Sons, New York, USA, vol 21. pp 139–180 check well for vol #.

Hummer K, Hancock JF (2009) Strawberry genomics: Botanical history, cultivation, traditional breeding and new technologies. In: KM Folta , SE Gardiner (eds) Genetics and Genomics of Rosaceae. Springer, New York, USA, pp 238–495.

Hummer KE, Nathewet P, Yanagi T (2009) Decaploidy in *Fragaria Iturupensis* (Rosaceae). Am J Bot 96: 713–716.

Ichijima K (1926) Cytological and genetic studies on *Fragaria*. Genetics 11: 590–603.

Igrashi I, Monma S, Fujino M, Okimura M, Okitsu S, Takada K, Nii T (1994) Breeding of new strawberry varieties "Ever Berry". Bull Natl Res Inst Veg Ornam Plants Tea Jpn Ser 9: 69–84.

Iwatsubo Y, Naruhashi N (1989) Karyotypes of three species of Fragaria (Rosaceae). Cytologia 54: 493–497.

James CM, Wilson F, Hadonou AM, Tobutt KR (2003) Isolation and characterization of polymorphic microsatellites in diploid strawberry (*Fragaria vesca* L.) for mapping, diversity studies and clone identification. Mol Ecol Notes 3: 171–173.

Korbin M, Kuras A, Zurawicz E (2002) Fruit plant germplasm characterisation using molecular markers generated in RAPD and ISSR-PCR. Cell Mol Biol Lett 7: 785–794.

Kunihisa M, Fukino N, Matsumoto S (2003) Development of cleavage amplified polymorphic sequence (CAPS) markers for identification of strawberry cultivars. Euphytica 134: 209–215.

Kunihisa M, Fukino N, Matsumoto S (2005) CAPS markers improved by cluster-specific amplification for identification of octoploid strawberry (*Fragaria x ananassa* Duch.) cultivars, and their disomic inheritance. Theor Appl Genet 110: 1410–1418.

Kuras A, Korbin M, Zurawicz E (2004) Comparison of suitability of RAPD and ISSR techniques for determination of strawberry (*Fragaria x ananassa* Duch.) relationship. Biotechnologia 2: 189–193.

Lerceteau-Köhler E, Guerin G, Laigret F, Denoyes-Rothan B (2003) Characterization of mixed disomic and polysomic inheritance in the octoploid strawberry (*Fragaria x ananassa*) using AFLP mapping. Theor Appl Genet 107: 619–628.

Lerceteau-Köhler E, Guerin G, Denoyes-Rothan B (2005) Identification of SCAR markers linked to *Rca 2* anthracnose resistance gene and their assessment in strawberry germ plasm. Theor Appl Genet 111: 862–870.

Longley A (1926) Chromosomes and their significance in strawberry classification. J Agri Res 32: 559–568.

Monma S, Okitsu S, Takada K (1990) Inheritance of the everbearing habit in strawberry. Bull Natl Res Inst Veg Ornam Plants Tea Jpn Ser 1: 21–30.

Nehra NS, Kartha KK, Stushnoff C (1991) Nuclear-DNA content and isozyme variation in relation to morphogenic potential of strawberry (*Fragaria X Ananassa*) callus-cultures. Can J Bot 69: 239–244.

Nier S, Simpson DW, Tobutt KR, Sargent DJ (2006) A genetic linkage map of an inter-specific diploid Fragaria BC1 mapping population and its comparison with the Fragaria reference map (FV X FN). J Hort Sci Biotechnol 81: 645–650.

Nyman M, Wallin A (1992) Improved culture technique for strawberry (*Fragaria x Ananassa* Duch) protoplasts and the determination of DNA content in protoplast derived plants. Plant Cell Tiss Org Cult 30: 127–133.

Ozkan H, Tuna M, Arumuganathan K (2003) Nonadditive changes in genome size during allopolyploidization in the wheat (*Aegilops-Triticum*) group. J Hered 94: 260–264.

Pontaroli A, Rogers R, Zhang Q, Davis T, Folta KM, SanMiguel P, Bennetzen J (2009) Gene content and distribution in the nuclear genome of *Fragaria vesca*. Plant Genome 2: 93–101.

Potter D, Luby JJ, Harrison RE (2000) Phylogenetic relationships among species of *Fragaria* (Rosaceae) inferred from non-coding nuclear and chloroplast DNA sequences. Syst Bot 25: 337–348.

Qu L, Hancock JF (2001) Detecting and mapping repulsion-phase linkage in polyploids with polysomic inheritance. Theor Appl Genet 103: 136–143.

Ripol MI, Churchill GA, da Silva JAG, Sorrells M (1999) Statistical aspects of genetic mapping in autopolyploids. Gene 235: 31–41.

Rousseau-Gueutin M, Lerceteau-Köhler E, Barrot L, Sargent DJ, Monfort A, Simpson D, Arús P, Guerin G, Denoyes-Rothan B (2008) Comparative genetic mapping between octoploid and diploid fragaria species reveals a high level of colinearity between their genomes and the essentially disomic behavior of the cultivated octoploid strawberry. Genetics 179: 2045–2060.

Rousseau-Gueutin M, Gaston A, Ainouche A, Ainouche ML, Olbricht K, Staudt G, Richard L, Denoyes-Rothan B (2009) Tracking the evolutionary history of polyploidy in Fragaria L. (strawberry): new insights from phylogenetic analyses of low-copy nuclear genes. Mol Phylogenet Evol 51: 515–530.

Rousseau-Gueutin M, Richard L, Le Dantec L, Caron H, Denoyes-Rothan B (2010) Development, mapping and transferability of *Fragaria* EST-SSRs within the Rosodae supertribe. Plant Breed 129: 1–8.

Rugienius R, Siksnianas T, Stanys V, Gelvonauskiene D, Bendokas V (2006) Use of RAPD and SCAR markers for identification of strawberry genotypes carrying red steele (*Phytophtora fragariae*) resistance gene *Rpf1*. Agron Res 4: 335–339.

Sánchez-Sevilla JF, Ariza DG, Amaya I, Botella MA, López-Aranda J, Valpuesta V (2009) Impact of plant breeding on the genetic diversity of cultivated strawberry as revealed by EST-derived SSR markers. J Am Soc Hort Sci pages volume??

Sargent DJ, Hadonou AM, Simpson DW (2003) Development and characterization of polymorphic microsatellite markers from *Fragaria viridis*, a wild diploid strawberry. Mol Ecol Notes 3: 550–552.

Sargent DJ, Davis TM, Tobutt KR, Wilkinson MJ, Battey NH, Simpson DW (2004) A genetic linkage map of microsatellite, gene-specific and morphological markers in diploid *Fragaria*. Theor Appl Genet 109: 1385–1391.

Sargent DJ, Battey NH, Wilkinson MJ, Simpson DW (2006a) The detection of QTL associated with vegetative and reproductive traits in diploid Fragaria: A preliminary study. Acta Hort 708: 471–474.

Sargent DJ, Clarke J, Simpson DW, Tobutt KR, Arús P, Monfort A, Vilanova S, Denoyes-Rothan B, Rousseau M, Folta KM, Bassil NV, Battey NH (2006b) An enhanced microsatellite map of diploid *Fragaria*. Theor Appl Genet 112: 1349–59.

Sargent DJ, Rys A, Nier S, Simpson DW, Tobutt KR (2007) The development and mapping of functional markers in *Fragaria* and their transferability and potential for mapping in other genera. Theoretical and Applied Genetics 114(2): 373–384.

Sargent DJ, Fernandez-Fernandez F, Ruiz-Roja JJ, Sutherland BG, Passey A, Whitehouse AB, Simpson DW (2009) A genetic linkage map of cultivated strawberry (*Fragaria x ananassa*) and its comparison to the diploid *Fragaria* reference map. Mol Breed 24: 293–30.

Senanayake YDA, Bringhurst RS (1967) Origin of *Fragaria* polyploids .I. cytological analysis. Am J Bot 54(2): 221–228.

Shaw DV (2003) Heterogeneity of segregation ratios from selfed progenies demonstrate polygenic inheritance for day neutrality in strawberry (*Fragaria xananassa* Duch.). J Am Soc Hort Sci 128: 504–507.

Shaw DV, Famula TR (2005) Complex segregation analysis of day-neutrality in domestic strawberry (*Fragaria x ananassa* Duch.). Euphytica 145: 331–338.

Sorrells ME (1992) Development and application of RFLPs in polyploids. Crop Sci 32: 1086–1091.

Stafne ET, Clark JR, Weber CA, Graham J, Lewers KS (2005) Simple sequence repeat (SSR) markers for genetic mapping of raspberry and blackberry. J Am Soc Hort Sci 130: 722–728.

Staudt G (1973) Fragaria iturupensis, eine neue Erdbeerart aus Ostasien. Willdenowia 7: 101–104.

Staudt G (1989) The species of *Fragaria*, their taxonomy and geographic distribution. Acta Hort 265: 23–33.

Staudt G (2003) Notes on Asiatic Species III: *Fragaria orientalis* Losinsk. and *Fragaria mandshurica* spec. nov. Bot Jahrb Syst.

Staudt G (2009) Strawberry biogeography, genetics and systematics. Acta Hort. 842(1): 71–83.

Sugimoto T, Tamaki K, Matsumoto J, Yamamoto Y, Shiwaku K, Watanabe K (2005) Detection of RAPD markers linked to the everbearing gene in Japanese cultivated strawberry. Plant Breed 124: 498–501.

Tyrka M, Dziadczyk P, Hortynski JA (2002) Simplified AFLP procedure as a tool for identification of strawberry cultivars and advanced breeding lines. Euphytica 125: 273–280.

Weebadde CK, Wang D, Finn CE, Lewers KS, Luby JJ, Bushakra J, Sjulin TM, Hancock JF (2008) Using a linkage mapping approach to identify QTL for day-neutrality in the octoploid strawberry. Plant Breed 127: 94–101.

Wendel JF (2000) Genome evolution in polyploids. Plant Mol Biol 42: 225–249.

Wu KK, Burnquist W, Sorrells ME, Tew TL, Moore PH, Tanksley SD (1992) The detection and estimation of linkage in polyploids using single-dose restriction fragments. Theor Appl Genet 83: 294–300.

Zorrilla-Fontanesi Y, Cabeza A, Torres AM, Botella MA, Valpuesta V, Monfort A, Sánchez-Sevilla JF, Amaya I (2010) Development and bin mapping of strawberry genic-SSRs in diploid *Fragaria* and their transferability across the Rosoideae subfamily. Mol. Breed. DOI 10.1007/s11032-010-9417-1.

Strawberry

Part 3: Structural and Functional Genomics

Janet P. Slovin[1]* and *Todd P. Michael*[2]

4.3.1 Introduction

This part of the strawberry chapter describes progress in strawberry genomics, proteomics and metabolomics, and summarizes some of the recent progress that has been made in strawberry gene discovery and characterization since the review by Folta and Davis (2006). Where appropriate, potential directions in the field will be discussed, highlighting some of the progress described at two somewhat recent meetings: The VI International Strawberry Symposium held in Huelva, Spain in March 2008; and The International Rosaceae Meeting held in Pichon, Chile, also in March 2008.

4.3.2 *Fragaria vesca*: A model Fragaria Species

The rapidly changing arena of strawberry genomics has recently been discussed in light of the burgeoning interest in, and the need for, reference plants for the Rosaceae family, which contains many important fruit, nut, ornamental and wood crops (Shulaev et al. 2008). Because of the diversity of family members, the Rosaceae research community has recognized the need for multiple reference species. One of the plants chosen is the diploid strawberry, *Fragaria vesca*. As has been discussed in the previous chapters, the cultivated strawberry, *F.* × *ananassa* is octoploid; a hybrid resulting from a cross between the native octoploids, *F. virginiana* and *F. chiloensis*. Commercial interests have dictated that until recently, research be focused

[1]Genetic Improvement of Fruit and Vegetables Laboratory, Henry A. Wallace Beltsville Agricultural Research Center, USDA/ARS Beltsville, MD.
[2]Waksman Institute, Rutgers University, Piscataway, NJ.
*Corresponding author: *janet.slovin@ars.usda.gov*

on the commercial octoploid, in particular on fruit development and quality, or on pest resistance.

Several groups recognized very early the utility of working with a diploid *Fragaria* rather than an octoploid (Battey et al. 1998; Davis and Pollard 1991; Slovin and Rabinowicz 2007; Uratsu et al. 1991). The diploid, *F. vesca*, can be considered an ideal model perennial plant because it is easy to propagate sexually and vegetatively as a small plant in the laboratory, it can produce many seeds per plant, it has a short seed to seed cycle, its genome size is small, and it is amenable to genetic transformation. *F. vesca* is widely found throughout the temperate zone and has a diverse germplasm base. In addition, *F. vesca* is a good ancestral candidate for the octoploid species.

As described in the previous chapter, a diploid *Fragaria* reference map is available and is rapidly being populated with SSR and gene-specific markers. The high-level of colinearity between the genomes of octoploid and diploid *Fragaria* provides an additional argument for the use of *F. vesca* as a research tool in studying the more complex octoploid (Rousseau-Gueutin et al. 2008). Various named and un-named genotypes of *F. vesca* are already being used for studies of gene function and biochemistry (e.g., Aharoni et al. 2004; de la Fuente et al. 2006; Osorio et al. 2008; Ruiz-Rojas et al. 2008). Adoption by the strawberry research community of one or a few largely homozygous, well-characterized *F. vesca* genotypes for genomics would make it easier to compare results among laboratories and would facilitate building upon a knowledge base. Inbred lines of *F. vesca* have been described (Battey et al. 1998; Slovin and Rabinowicz 2007) and seeds of lines 5AF7 and Ruegen F7-4 are available to the community. Ruegen F7-4 is non-runnering, and has red achenes on a red-receptacle, which has a grape-like aroma. 5AF7 produces sweet smelling pale yellow fruit with tan achenes, and the plant is also non-runnering. An effort is underway to obtain an efficiently transformable inbred line from Hawaii-4. Hawaii-4 fruit is similar to that of 5AF7, however the plant produces abundant runners. All three *F. vesca* genotypes are day-neutral and will produce fruit throughout the year in the greenhouse.

4.3.3 *Fragaria* Structural Genomics

The number of ESTs available in the public domain for all species of *Fragaria* as of March 2010 is 58,622 (Table 4.3-1). Of these, 47,743 are from *F. vesca*, with the majority being from abiotically stressed tissues of the *F. vesca* line Hawaii-4 (Table 4.3-2). The Hawaii-4 sequences were obtained, with funding from the CSREES, USDA National Research Initiative, to increase the diversity of ESTs for *Fragaria* because, as might be surmised from a list of known *Fragaria* cDNA libraries in Table 4.3-2, about 75% of the nucleotide sequence data available for strawberry came from *F.* × *ananassa* fruit transcripts at various stages of development.

Although several strawberry microarray experiments have been reported, only 10,830 EST sequences from the commercial strawberry, *F.* × *ananassa* are publicly available in dbEST (Table 4.3-1), presumably for proprietary reasons. Of the 1113 *F.* × *ananassa* sequences in the NCBI nucleotide database, 556 sequences are from patents, and it is quite plausible that there is a large amount of additional sequence data in private databases. Of the remaining *F.* ×*ananassa* sequences in the nucleotide database, 95 are microsatellite clones, and 102 are patented sequences designated "Method for distinguishing strawberry variety". Some of these sequences are as short as four nucleotides. As of March 2009, only 171 full-length cDNA sequences for *F.* ×*ananassa* were present in the GenBank, and only 31 entries were annotated genomic sequence, containing introns and/or promoter regions.

Only 10 full-length *F. vesca* coding sequences were available as of March 2009. A new initiative to sequence and publish the transcriptome of *F. vesca* line Hawaii-4 has been launched (K. Folta and K Mockaitis, pers. comm.) and the resulting information will be extremely useful for many applications

Table 4.3-1 *Fragaria* nucleotide sequences in the GenBank

Species	# in dbEST	# in dbNUC	dbNUC tRNA/ rRNA[1]	GSS[2]
F. vesca				0
subsp.				
vesca	47,743	261[3]	31	
americana	0	60[4]	2	
bracteata	0	3	0	
Other diploid species	0	180[5]	91	0
F. orientalis	0	16	10	0
F. moschata	0	17	11	0
F. chiloensis	0	44	21	0
F. virginiana	0	49	28	2
F. ×*ananassa*	10,830	1113[6]	14	15
Total[7]	58,622	1,839	238	17

[1]Number of sequences in the GenBank nucleotide database (dbNUC) that encode *Fragaria* tRNA or rRNA.
[2]The Genome Survey Sequence (GSS) records are similar to ESTs except that sequences are of genomic origin.
[3]Includes 103 microsatellite clones, 58 cds, and 25 patented sequences.
[4]Includes 24 complete fosmid sequences and 21 working draft fosmid sequences.
[5]Includes 63 microsatellite clones and 11 cds.
[6]Includes 95 microsatellite clones and 556 sequences from patents, of which 105 sequences relate to methods for distinguishing strawberry varieties.
[7]As of March 2010.

Table 4.3-2 Fragaria cDNA libraries used for ESTs in GenBank.

Library Name	Species	Tissue	Developmental Stage or Treatment	Number of ESTs	Contact/email
F. × *ananassa* Camarosa crown	× *ananassa*	crown	Uninfected	16	Caballero, J.L. bb1carej@uco.es
F. × *ananassa* Andana crown	× *ananassa*	crown	Infection with *Colletotrichum*	27	Caballero, J.L. bb1carej@uco.es
Colletotrichum infected leaf library	× *ananassa*	leaf	Infection with *Colletotrichum*	12	Tonello, U. M. ursulat@unt.edu.ar
Strawberry pechika leaf cDNA	× *ananassa*	leaf	Wild-type mature	22	Adacho, Y adachi-yo632@pref.miyagi.jp
Strawberry pechika mutant leaf cDNA	× *ananassa*	leaf	Mutant	1	Adacho, Y adachi-yo632@pref.miyagi.jp
'Strawberry Festival' cDNA	× *ananassa*	whole plant	24 h post spray with 1.0mM Salicylic Acid, 4.0mM SA soak	1,511	Folta, K. kfolta@ifas.ufl.edu
KM-1	× *ananassa*	receptacle	Pink	23	Manning, K. ken.manning@hri.ac.uk
Strawberry UniZapXR cDNA library	× *ananassa*	achenes and receptacles	Red ripe	36	Aharoni, A. asaph.aharoni@weizmann.ac.il
LambdaZap Express library	× *ananassa*	fruit	Red ripe	3,753	Perrotta, G. gaetano.perrotta@trisaia.enea.it
UMA library M2	× *ananassa*	achenes	Green F3	1,238	Sanchez-Sevilla, J. F. josef.sanchez@juntadeandalucia.es
UMA Library M1	× *ananassa*	receptacles	Green F3	1,863	Sanchez-Sevilla, J. F. josef.sanchez@juntadeandalucia.es
Elsanta fruit library L1	× *ananassa*	achenes and receptacles	Red ripe ethylene treated	345	Sanchez-Sevilla, J. F. josef.sanchez@juntadeandalucia.es
Subtractive (red-green) fruit library	× *ananassa*	achenes and receptacles	Red fruit with green fruit subtracted	1	Sanchez-Sevilla, J. F. josef.sanchez@juntadeandalucia.es

Table 4.3-2 contd...

Table 4.3-2 contd....

Library Name	Species	Tissue	Developmental Stage or Treatment	Number of ESTs	Contact/email
Chandler red fruit UCO library C3	× *ananassa*	achenes and receptacles	Red fruit	1,100	Munoz-Blanco, J. bb1mub1j@uco.es
F. × *ananassa*	× *ananassa*	leaf and root	Organic and inorganic fertilizer	14[2]	Zaccagnino, N. nadia.zaccagnino@unibas.it
F. × *ananassa*	× *ananassa*	unknown	Unknown	103	Denoyes-Rothan, B. denoyes@bordeaux.inra.fr
'Strawberry Festival' flower library	× *ananassa*	sexual reproductive	early bud to senescent flower	737	Folta, K. kfolta@ifas.ufl.edu
Yellow Wonder flower buds	vesca	reproductive	Closed buds	2,716	Davis, T. tom.davis@unh.edu
Yellow Wonder heat stressed seedlings[1]	vesca	aseptic seedlings	37°C or 42°C for 1-4 h	1,306	Slovin, J. P. janet.slovin@ars.usda.gov
Hawaii-4 heat stressed[1]	vesca	whole plant	37°C or 42°C for 1-4 h	8,885	Slovin, J. P. janet.slovin@ars.usda.gov
Hawaii-4 cold stressed[1]	vesca	whole plant	4-5°C for 0.25-24 h	9,639	Slovin, J. P. janet.slovin@ars.usda.gov
Hawaii-4 drought stressed[1]	vesca	whole plant	Water withheld for 1-5 wilt cycles	6,846	Slovin, J.P. janet.slovin@ars.usda.gov
Hawaii-4 salt stressed[1]	vesca	aseptic seedlings and whole plant	25-75 mM NaCl	8,577	Slovin, J. P. janet.slovin@ars.usda.gov
Hawaii-4 heat and salt stressed[1]	vesca	whole plant	Combined heat and salt stress	7,483	Slovin, J. P. janet.slovin@ars.usda.gov
F. vesca	vesca	leaf and root	Organic and inorganic fertilizer	29[2]	Zaccagnino, N. nadia.zaccagnino@unibas.it
F. vesca subsp. vesca	vesca	unknown	Unknown	108	Denoyes-Rothan, B. denoyes@bordeaux.inra.fr
Suppress-subtract hyb lib SD PI551792	vesca	shoot tips	Long daylength	968	Mouhu, K. katriina.mouhu@helsinki.fi
Suppress-subtract hyb lib EB PI551507 (Baron Solemacher)	vesca	shoot tips	Long daylength	1,184	Mouhu, K. katriina.mouhu@helsinki.fi

[1]Library constructed in Gateway® vectors. [2]From differential display.

including microarray design and molecular marker development, as well as for genome annotation. At the present time, no such public initiative has been discussed in the US for octoploid strawberry. No ESTs are available for either of the two immediate progenitors of the commercial hybrid, *F. chiloensis* and *F. virginiana*, and most of the nucleotide database sequences for these species are tRNA or rRNA sequences used for phylogenetic studies (Table 4.3-1). Only 11 protein coding region sequences from other diploids are available.

Three assemblies of *Fragaria* ESTs are in the public domain (Table 4.3-3). One assembly can be found within the Genome Database for the Rosaceae (GDR) (Jung et al. 2008) and it is a combined assembly of all the publicly available diploid and octoploid *Fragaria* EST sequences. A total of 13,896 putative unigenes were identified in this assembly. The GDR also features a Rosaceae unigene set that is a combined assembly of the EST assemblies from several Rosaceae family members. The Rosaceae unigenes set is highly skewed toward *Malus* however, because over > 336,000 ESTs are currently available from various *Malus* species, as compared to the approximately 50,000 from *Fragaria*. Two other *Fragaria* assemblies (Table 4.3-3) can be found at the TIGR Plant Transcript Assembly website (Childs et al. 2007). The TIGR transcript assemblies include sequences from the NCBI nucleotide database as well as ESTs. Sequences from *F. vesca* were assembled into 4,825 contigs and 8,624 singlets. Sequences from *F.* × *ananassa* assembled into 358 contigs and 4,776 singlets.

Most of the *F. vesca* libraries in Table 4.3-2 were made using pooled RNA samples from whole plants. Recently, diploid strawberry homologs for *Arabidopsis* flowering time gene sequences were identified in shoot tip subtractive hybridization libraries from short day and everbearing varieties (Mouhu et al. 2009). Although laser capture microdissection (LCM) was

Table 4.3-3 *Fragaria* Sequence Assemblies.

	GDR	*F. vesca* TIGR TA	*F.* × *ananassa* TIGR TA
# sequences used for assembly	49,132[1]	45,085	5,585[2]
# of contigs or assemblies	5,582	4,825	358
# of singlets	8,314	8,624	4,776
# putative unigenes	13,896	13,449	5,134
Assembly programs	CAP3[4]	TGICL[3] Megablast[5] CAP3	TGICL Megablast CAP3

[1]Number resulting from quality filtering of 50,882 total available ESTs.
[2]Includes 258 full-length or partial cDNA sequences.
[3]Pertea et al. 2003 Bioinformatics 19: 651–652.
[4]Thompson et al. 1994 Nucl Acids Res 22: 4673–4680.
[5]Zhang et al. 2000 J Comput Biol 7: 203–214.

used to obtain strawberry fruit tissue for gene expression analysis (Raab et al. 2006), there have been no reported tissue specific cDNA libraries constructed using this technique. In combination with new sequencing technology, LCM should greatly expand the strawberry transcriptome to those sequences that are only expressed in a small number of specialized cells, and could potentially reveal genes that are unique to *Fragaria*. There are no public initiatives to specifically obtain a collection of full-length cDNA sequences from *Fragaria*, however, several of the cDNA libraries listed in Table 4.3-2 were constructed so as to maximize full length clones.

As described in the previous part of this strawberry chapter, *Fragaria* ESTs have been screened to develop gene-specific simple sequence repeat (EST-SSR) markers (Bassil et al. 2006b; Bassil et al. 2006c; Gil-Ariza et al. 2006b; Keniry et al. 2006; Lewers et al. 2005; Sargent et al. 2007b), some of which have now been mapped to the *Fragaria* reference map using "selective" or "bin" mapping (Sargent et al. 2008). Most of these SSRs are from fruit library ESTs, which until recently were the predominant sequences available. The *F. vesca* Hawaii-4 stress library sequences are currently being analyzed for SSRs (J. Slovin and P. Rabinowicz, unpubl), and these should greatly increase the number of gene-specific markers on the strawberry maps. Several gene specific markers have also been developed for genes that are highly expressed in response to elevated temperature in *F. vesca* 5AF7 by exploiting polymorphisms in intron length and sequence (J. Slovin, unpubl). In the near future, the enormous amount of sequence data generated by ongoing *F. vesca* genome and transcriptome sequencing projects will greatly increase the number of identified genes on *Fragaria* maps.

4.3.4 Genome Sequencing Initiatives

Collaboration among the Davis (University of New Hampshire), Folta (University of Florida), Bennetzen (University of Georgia) and SanMiguel (Purdue Univerity) laboratories resulted in the sequencing of almost 2 Mb, or roughly 1%, of the *F. vesca* genome, which has been estimated to be about 200 Mb (Folta and Davis 2006). Sequence was obtained from a fosmid library constructed using DNA from the *F. vesca* subsp. *americana* var. "Pawtuckaway", which is one of the mapping parents used for a new, intraspecific, diploid map (Davis et al. 2007). Preliminary analyses of sequence from 50 of these fosmids were described in a report by Davis et al. at the International Rosaceae Meeting in Chile, 2008 and a more complete analysis has now been published describing 12 completely sequenced fosmid insert representing ~ 1.0 Mb of nuclear genomic DNA (Pontaroli et al. 2009). From this data it could be predicted that at least 16% of the *F. vesca* subsp. *americana* is comprised of transposable elements, with the most abundant element being long terminal repeat retrotransposons. The pattern of gene distribution in *F. vesca* is more similar to *Arabidopsis* than to grasses, with an average of one gene per 6 kb, although this pattern may be biased in favor of gene-rich regions because approximately 50% of the fosmids sequences were selected based on their containing a "gene of interest". For a long time this has been the largest publicly available genomic resource for all members of the Rosaceae family (Shulaev et al. 2008). However, a draft assembled and annotated genome of peach was made available in April 2010 by the International Peach Genome Initiative, and it is anticipated that a draft genome sequence of a fourth generation inbred line of *F. vesca* Hawaii 4 will be available before this chapter is published.

The genome sizes of several *F. vesca* genotypes have recently been re-evaluated using flow cytometry (T. Michael, unpubl; Figure. 4.3.1). These measurements indicated that the *F. vesca* inbred line 5AF7 could have a slightly larger genome than the fourth generation inbred of Hawaii-4. A consortium was established at the Plant and Animal Genome 2008 conference in San Diego, California, to sequence the genome of Hawaii-4 using Roche 454 technology, and it is this sequence that is forthcoming.

The availability of the *F. vesca* genome sequence will have an enormous impact on our understanding of how the genomes of the octoploid strawberries are structured and how they evolved. The relatively small amount of *F. vesca* genomic sequence that does exist has already resulted in substantial new insights (Davis et al. 2008a). Both Copia and Gypsy-like retrotransposon related sequences were found. SSRs occurred on average

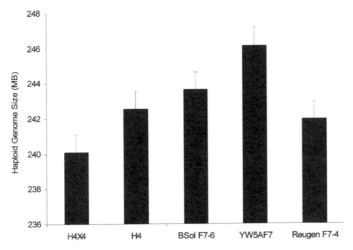

Figure 4.3-1 Haploid genome sizes for various *Fragaria vesca* genotypes. H 4x4 is a fourth generation inbred line of Hawaii-4. H4 represents Hawaii-4, BSol F7-6 is an advanced inbred line of 'Baron Solemacher', YW5AF7 is an inbred yellow wonder selection and Reugen F7-4 is an inbred line from the 'Reugen' genotype.

every 6 kb. Regions of microcolinearity were found between the strawberry and other sequenced genomes including *Arabidopsis, Vitis,* and *Populus* (Davis et al. 2008a; Davis et al. 2008b; Folta and Davis 2006), increasing the possibility that these already sequenced genomes could serve as scaffolds for assembly of the *F. vesca* genomic sequence.

An additional benefit from the fosmid sequencing project has been the ability to detect "gene pair loci" for mapping (Davis et al. 2007). As described in part 2 of this chapter, resequencing of intergenic regions using primers to conserved coding regions in the flanking genes, has implicated the Japanese diploid, *F. iinumae* as one of the genome donors to the octoploid strawberries (Davis et al. 2008a). In two *F. virginiana* accessions, the allelic variation found in the intergenic region between *gRGA1*, a resistance-like gene, and *Subtilase*, a putative serine protease gene, indicated that three or more distinct genomes are represented in the octoploid strawberries, and that an ancestor of *F. iinumae* is likely have contributed to the mix (Davis et al. 2008a).

A recent report on SNP analysis of the sorbitol transferase gene in *F. × ananassa* supports the view that there were three or more ancestral progenitors to today's octoploids (Kaur et al. 2008).

4.3.5 Functional Genomics: Transcriptomics

The forthcoming transcriptome sequences from *F. vesca* Hawaii-4 will be invaluable for annotating the *F. vesca* genome, as well as for providing candidate genes for gene function studies. The function(s) of such candidate genes can be studied using gene knockout approaches such as insertion mutagenesis and RNAi, or by overexpression in *Fragaria* or heterologous systems such as *Arabidopsis* or tobacco. In combination with genetics, these approaches have been highly successful in elucidating gene function in *Arabidopsis*, and these approaches are beginning to shed light on strawberry gene function as well.

Several excellent reviews of the many reports on strawberry tissue culture and transformation have recently been published (Debnath and Teixeira da Silva 2007; Folta and Davis 2007; Folta and Dhingra 2007). Strawberry appears to be amenable to many tissue culture techniques. Reports exist of protoplast fusion (Wallin 1997) and of anther culture resulting in haploid plants (Owen and Miller 1996). High frequency somatic embryogenesis from strawberry callus has now been achieved using 2,4-D, BAP, and proline with continuous darkness (Biswas et al. 2007). Large scale production of *Agrobacterium*—mediated transgenic strawberries can be achieved using a temporary immersion bioreactor system (Hanhineva and Kärenlampi 2007).

In addition to *Agrobacterium*, particle bombardment can be used to introduce a gene into strawberry for stable transformation (Wang et al. 2004), and for transient gene expression (Spolaore et al. 2001) or promoter analysis (Agius et al. 2003). Coating gold particles with *Agrobacterium* improved the transformation efficiency of leaf explants using the biolistics approach (Cordero de Mesa et al. 2000). Using a needle and syringe to inject *Agrobacterium*, Hoffman et al. (2006) successfully transfected fruit of *F.* × *ananassa* with an RNAi-induced silencing construct for chalcone synthase (*CHS*), resulting in reduced levels of *CHS* mRNA.

Despite the number of positive reports, the success of any tissue culture or transformation technique with strawberry appears to be genotype specific. Efficient transformation and rapid regeneration are highly desirable for gene function studies, so two groups have screened for lines that exhibit these properties (Folta et al. 2006; Oosumi et al. 2006). The small diploid *F. vesca* Hawaii-4 and the octoploids, *F.* × *ananassa* line LF-9 and cultivar Selva, are currently being exploited to study strawberry gene function using forward and reverse genetic approaches (Folta and Davis 2007; Mezzetti and Constantini 2006). It is anticipated that the availability of transformable inbred lines of *F.* vesca will further facilitate gene function studies (Slovin et al. 2009). Easily transformed heterologous plant systems such as *Arabidopsis* have also been used to shed light on strawberry gene function. *Arabidopsis* plants transformed with genes that had been identified as being novel to the Rosaceae exhibited a number of noteworthy phenotypes, including failure to produce trichomes, assymetrical rosettes, and late or early flowering (Folta and Davis 2007).

A population of T-DNA insertion lines for gene tagging has been generated using Hawaii-4 (Oosumi et al. 2006), and the sequences of T-DNA flanking regions in 28 mutant lines have already been deposited in the GenBank. Although it is estimated that at least 255,000 individual lines will need to be transformed to mutate all the genes in the diploid, the small plant size and ease of transformation of *F. vesca* make it possible to do high throughput generation of T-DNA mutants with these plants (Oosumi et al. 2006). In a report at the International Rosaceae Meeting in Chile, 2008, V. Shulaev described mapping 12 of these T-DNA insertion events to the diploid reference map (Sargent et al. 2007) using SNPs discovered in the flanking sequences.

4.3.5.1 *Fruit Growth and Quality*

As can be seen from previous reviews (Folta and Davis 2006; Hokanson and Maas 2001) the emphasis for most of the recent research on gene expression and function in strawberry has focused on some aspect of fruit production, including fruit development, flavor and aroma development, nutritional attributes, fruit firmness, and ripening. There are now sufficient numbers of groups working on strawberry so that standards for describing flower and fruit developmental stages could be set by the research community in order to coordinate data. The 10 stages from closed bud to ripe fruit shown by Smith (2007) expands on the seven shown in de la Fuente (2006) for example, and describe the major developmental stages, including stages in green fruit growth that span the changes from early cell division to primarily cell expansion. These could be used as the standard if the additional stage, overripe (senescing) red fruit is added (Table 4.3-4). However, these 11 stages may not even fully take into account the developmental changes occurring in the achene. The achene is the true botanical strawberry fruit, and produces the hormone auxin during specific stages of its development. This could have a profound effect on the developmental stage and physiological character of the underlying receptacle tissue. Additional details will undoubtedly need to be added to Table 4.3-4 as more information becomes available about the molecular and metabolic events occurring in the receptacle and achenes during fruit growth and senescence.

Of the approximately 800 *F.* × *ananassa* sequences currently available in the GenBank nucleotide database that are not designated molecular markers or non-protein coding (Table 4.3-1), about 75% can be directly related to fruit quality or development, including for example sequences

Table 4.3-4 Stages in fruit development.

Stage #	Description	Notes
1	Mature closed bud	
2	Opening bud, white petals visible	Anthers maturing
3	Fully open flower	Highly susceptible to anthracnose[1]
4	Pollinated flower, petals dropped	
5	Green fruit, < 25% of mature fruit size[2]	Achenes producing auxin[3]
6	Expanding green fruit, 25–75% mature fruit size	
7	Green fruit > 75% of the size of mature fruit	
8	White fruit	Achenes stop producing auxin[3]
9	Receptacle pink	Highly susceptible to anthracnose[1]
10	Mature red fruit	Highly susceptible to anthracnose[1]
11	Overripe (senescing) red fruit	

[1](Smith 2007).
[2] Mature fruit size can be influenced by the position of the fruit on the inflorescence and by the number of inflorescences on the plant.
[3](Archbold and Dennis 1984; Given et al. 1988).

encoding several glucosyltransferases (Griesser et al. 2008a; Griesser et al. 2008b; Landmann et al. 2007); pectin methyl esterases (Osorio et al. 2008; Dray and van Cutsem, unpublished) and pectate lyase (Santiago-Domenech et al. 2008); polygalacturonase (Villarreal et al. 2008), leucoanthocyanidin reductase and other flavonoid and proanthocyanidin biosynthesis enzymes (Almeida et al. 2007), an allergen (Musidlowska-Persson et al. 2007), and an aquaporin subtype (Mut et al. 2008). Investigations of the expression patterns and functioning of genes expressed in strawberry fruit have benefitted greatly from microarray technology (reviewed in Folta and Davis 2006). Using custom arrays, transcripts associated with ripening (Schwab et al. 2001), stress and auxin-induced gene expression (Aharoni et al. 2002), volatile production (Aharoni et al. 2004), fruit development (Aharoni and O'Connell 2002), and flavor (Aharoni et al. 2000a) have been identified. These arrays are not in the public domain however, nor are most of the sequences used to design the probes for these arrays.

In a recent set of fruit quality related microarray analyses, a custom cDNA array of more than 5,600 probes—representing 1,811 fruit non-public ESTs from the cultivar Queen Elisa (QE), 49 ripening related sequences available from public databases, and 10 negative controls—was used for comparative transcription profiling of red ripe fruit from five *F. × ananassa* genotypes (Carbone et al. 2006). These five genotypes were also evaluated and compared for firmness, soluble solid content (Brix), titratable acidity, and plant yield. Transcriptome profiling was integrated with metabolic profiling of the volatiles emitted by these five genotypes, and the data from QE was used as the reference. As would be expected from earlier studies, transcripts encoding alcohol acyltransferase (AAT) were more highly up regulated in the genotypes exhibiting the highest concentration of esters by mass spectral analysis of headspace volatiles. Expression levels of genes encoding enzymes involved in vascularization and cell wall structure also increased during ripening. In a comprehensive study of biochemistry and gene expression related to flavonoid and proanthocyanin biosynthesis during fruit development, again the expected pathway genes showed elevated expression levels prior to and during fruit color accumulation (Almeida et al. 2007).

Glycosyltransferases perform a mulititude of roles in cells, changing the function or localization of a wide range of molecules, including plant hormones and xenobiotics, as well as proteins. The glucosyltransferases found in strawberry receptacles and achenes have been the subject of several recent papers (Griesser et al. 2008a; Griesser et al. 2008b; Landmann et al. 2007). *FaGT1* encodes an enzyme responsible for formation of the first stable intermediate in the anthocyanin pathway. Expression of *FaGT1* detected in turning and red ripe fruit, and was negatively regulated by auxin (Griesser et al. 2008a). *FaGT2* encodes an enzyme that catalyzes the formation of glucose esters of cinnamic and p-coumaric acid in fruit, and has broad substrate specificity toward other benzoic acid derivatives that occur naturally in plants as well as xenobiotics (Landmann et al. 2007). *FaGT6* and *FaGT7* also exhibit broad substrate tolerance (Griesser et al. 2008b). Their negative expression patterns in response to auxin, and up-regulation in response to salicylic acid, suggest multiple roles for these glucosyltransferases in fruit ripening and in detoxification. These genes are strongly expressed in achenes of young green fruit, where considerable levels of flavonol glucosides were also found (Griesser et al. 2008b).

The postharvest shelf life of strawberries is short due to fruit softening, which increases susceptibility to fungal infection. Post harvest heat treatments can be used with intact fruits (Lurie 1998) and even lightly processed fruits (Garcia and Barrett 2000) to delay softening. Heat treatment (45°C, 3 hour in air) of white stage *F. × ananassa* cv. Selva fruit resulted in short-term reduced expression of several genes encoding enzymes known to be involved in cell wall metabolism, including endoglucanase, β-xylosidase, polygalacturonase, and a fruit specific expansin (Martínez and Civello 2008). Polygalacturonase activity and expression appears to be correlated with the level of fruit firmness during ripening (Villarreal et al. 2008), whereas endo-[beta]-(1,4)-glucanase expression does not (Palomer et al. 2006). Although heat-treated fruit stayed firmer than non-treated fruit for over 24 hours, by 3 days of storage at 20°C there was no difference between treated and non-treated samples. It remains unclear whether heat treatments will be useful for delaying softening and maintaining fruit firmness in other cultivars, or once the fruit have reached full red stage, when color, taste, and aroma are the most desirable for consumers.

Using antisense technology, pectate lyase activity had been implicated in strawberry fruit softening (Jiménez-Bermudez et al. 2002), and this has recently been substantiated by biochemical and microscopic analyses of transgenic fruit (Santiago-Domenech et al. 2008). Differences in firmness between control and antisense transgenic fruit could be seen as early as the white stage, and microscopy indicated an inhibition of the breakdown of the middle lamella in red ripe transgenic fruit. Altogether, these studies on fruit softening indicate a significant role for pectin breakdown in strawberry ripening and senescence.

Additional aspects of fruit firmness might include the development of the vasculature, as evidenced by microarray data (Aharoni et al. 2002a), and the movement of water and solutes into and out of the fruit. The expression patterns throughout development and in response to auxin of a recently identified fruit specific plasma membrane aquaporin suggest that it may be responsible for water influx or efflux at the plasma membrane in ripe fruit, contributing to the maintenance or loss of fruit cell turgor (Mut et al. 2008).

A *F.* × *ananassa* gene, *FaGAST*, encoding a small protein with a cysteine-rich C-terminal domain characteristic of the *GAST* gene family, was identified by virtue of its being differentially expressed during development of strawberry fruit (de la Fuente et al. 2006). Members of this family have been shown to have heterogeneous functions, although many of them are induced by gibberellin (GA). *FaGAST* is expressed most prominently in older green stage and red ripe stage fruit, and is expressed in roots at the same level as in red ripe fruit. Heterologous expression of *FaGAST* in *F. vesca* and *Arabidopsis* resulted in similar phenotypes; reduced growth and delayed flowering. While the specific function of this gene in fruit growth was not determined, the data presented indicate a role for *FaGAST* in arrest of cell elongation (de la Fuente et al. 2006). Gibberellins have been implicated in strawberry runner development and elongation (Pankov 1992). *Fragaria* genes encoding enzymes involved in GA biosynthesis and GA signaling have been identified (Hytönen et al. 2009), however the role(s) of GAs in strawberry fruit or roots has not yet been addressed.

Strawberry fruit can elicit an allergic reaction in some people. Recently, genes for a strawberry allergen (Fra a 1) were cloned from several cultivars using primers designed from a protein sequence obtained by mass-spectrometry (Musidlowska-Persson et al. 2007). Five Fra a 1 protein isoforms could be predicted from 30 different clones, some matching the variations detected by mass-spectrometry (Hjernø et al. 2006). Recombinantly expressed Fra a 1 protein cross-reacted with antisera against apple and birch pollen homologs. The identification of such allergen genes could make it possible to eliminate allergens from the fruit, much as was done with soybean (Herman et al. 2003).

4.3.5.2 Flowering/Development

The transition from vegetative to reproductive growth is critical for strawberry fruit production, and equally critical for production of the runner plants required by the annual production schemes widely practiced today. The basic biological questions concerning how this transition is regulated are intriguing for researchers, and equally important for growers. The physiology of this transition has been studied for many years, revealing a complex interaction of genotype, nutrition, hormones, and light (Darnell et al. 2003). Only recently have the underlying molecular mechanisms for this transition begun to be studied in strawberry.

Two SSR markers have been mapped to within 2.2 cM of the seasonality locus in *F. vesca* (Albani et al. 2004; Cekic et al. 2001) and two QTLs, associated with the reproductive traits flower diameter and receptacle diameter, cluster in this same region of linkage group VI on the diploid *Fragaria* reference map (Sargent et al. 2006; Sargent et al. 2007). Additional QTLs associated with reproductive traits, including first flowering date and berry weight, were mapped to other linkage groups (Sargent et al. 2006). Identification of candidate genes for these traits awaits the linkage of the genetic map with the forthcoming *F. vesca* genome sequence and its annotation.

When the strawberry shoot apex switches from vegetative to reproductive stage there is a distinctive change from flat to dome shape (Durner and Poling 1985). In the June-bearing *F. × ananassa* cultivar Tochiotome, this morphological change was not seen until about 8 days after the central zone in the shoot apex begins to express the Histone H4 gene, which is considered to be an indication of cell division and an early indicator of floral induction (Kurokura et al. 2006). The signal for the shoot apex to initiate this cell division in the central zone is unknown in strawberry, although it is anticipated that many of the genes involved in floral induction and meristem transition in *F. × ananassa* will have already been identified in *Arabidopsis*.

CONSTANS (CO), a protein that has been shown to regulate flowering in *Arabidopsis*, has been identified in *Fragaria* as a single copy gene member of a small gene family that, as in *Arabidopsis*, accumulates and decays over a daily light/dark cycle (Stewart et al. 2007). Unlike what has been observed in other plants (Suarez-Lopez et al. 2001), where higher levels of *CO* mRNA occur at dawn and again at dusk, the *Fragaria* expression pattern exhibits a single peak at dawn, in both short-day and day-neutral cultivars (Stewart et al. 2007). This atypical expression pattern in *F. × ananassa* is being further investigated using RNAi and overexpression constructs to transform both octoploid and diploid strawberry (K. Folta, pers. comm.).

4.3.5.3 Disease/Pest Resistance

The research focus in disease and pest resistance has been on determining resistance mechanisms, and on using biotechnology to increase resistance. Strawberries are susceptible to infectious attack by insects, fungi, bacteria, viruses, mycoplasma-like organisms, and nematodes. The *Compendium of Strawberry Diseases*, published by The American Phytopathological Society (Maas 1984), is a good reference for information about strawberry diseases. This reference contains images of affected plant material, illustrates some of the effects of abiotic stresses on strawberry plants, and contains images of major pests. A recent report on strawberry flower blight and fruit rot caused by *Colletotrichum fragariae* and *C. acutatum* stands as an cautionary study, clearly pointing out that developmental stage and environment, in addition to genotype, can significantly affect susceptibility to attack by disease and pests (Smith 2007).

A large number of genes involved in the response of strawberry to *Colletotrichum* infection were identified in subtracted cDNA libraries using mRNA from the infected and non-infected crowns of two cultivars with differential susceptibility to infection (Casado-Diaz et al. 2006). Many of the identified genes showed sequence similarity to known plant defense genes such as LRR receptor-like proteins. Analysis of the expression of several of these genes indicated that, in some cases, fruit tissue and crown tissue can respond differently to infection. Of the genes tested, all exhibited a cultivar dependent pattern of expression, with the more resistant cultivar, Andana, exhibiting higher or more rapid expression of known defense genes than the susceptible cultivar, Camarosa (Casado-Diaz et al. 2006). Of particular interest is the finding that known pathogen response genes such as peroxidase, γ-thionin, NB-LRR proteins, chitanase and a β-1,3-glucanase were repressed in some tissues in response to pathogen infection (Casado-Diaz et al. 2006). Transformation of cultivar Camarosa with a β-1,3-glucanase gene from the fungus *Trichoderma* showed promise for increasing its resistance to *C. acutatum* (Mercado et al. 2007).

Two new β-1,3-glucanase genes have been identified in *F. × ananassa* (Shi et al. 2006). Both of these genes were induced by infection with *Colletotrichum*, with induction being higher upon infection with *C. fragariae* than with *C. acutatum*. Expression analysis also showed that these genes were regulated differently in all organs tested, and in response to infection. Neither of these genes shows sequence similarity to the β-1,3-glucanase identified by Casado-Diaz et al. (2006).

The *Pto* gene from tomato belongs to a family of plant disease resistance (*R*) genes that encode protein kinases. *Pto*, a serine-threonine kinase (STK), interacts with avirulence proteins from the pathogen resulting in a hypersensitive response (HR) characterized by rapid localized cell death (Kim et al. 2002.). *Pto*-like genes are being studied in wild and cultivated strawberry because of their potential as candidates for increasing broad-spectrum resistance (Martínez Zamora et al. 2004, 2008). STK sequences obtained from cultivated strawberry, *F. chiloensis*, and *Potentilla tucumanensis* could be grouped into five classes. The sequences from *F. vesca*, together with three sequences from *Potentilla* and one from *F. × ananassa* cv. Camarosa, grouped into two separate, distinct, classes. These two classes were determined to be non-*Pto*-like STKs because the sequences lacked one of the autophosphorylation sites required for elicitation of the HR, and were more similar to lectin receptor-like kinases. The function of all of these strawberry STKs in disease resistance remains to be determined.

Components of the plant cell wall, which acts as a physical barrier to pathogens, can also function as elicitors of plant cell defense responses. When a previously isolated *F. × ananassa* pectin methylesterase gene (*FaPE1*) was exogenously expressed in *F. vesca*, transgenic fruit exhibited increased resistance to infection with *Botrytis cinerea* (Osorio et al. 2008). Expression of the gene encoding pathogenesis related protein PR5 was elevated over wild type in the resistant transformants, as were the levels of salicylic acid. A decrease in the size of a specific class of pectin breakdown fragments, $[1 \rightarrow 4]$-α-linked oligogalacturonides (OGA), was found in the transformants. The degree of esterification of the OGA in the plant appears to determine the ultimate size of these fragments. Purified OGA from the transformants induced expression of *PR5* to a much higher level than OGA from wild type strawberry (Osorio et al. 2008). Interestingly, this same purified OGA did not elicit *PR5* expression in tobacco (*Nicotiana benthamiana*).

Using PCR with degenerate primers and a genome walking approach, a gene encoding an osmotin-like protein (OLP) was isolated from *F. × ananassa* (Zhang and Shih 2007). The osmotin and osmotin-like proteins belong to the thaumatin-like or PR5 family of proteins that have been implicated in abiotic stress responses and in responses to fungal pathogens, as well as a number of developmental processes. The strawberry OLP gene (*FaOLP2*) is expressed in leaves, roots, and fruit, but is most highly expressed under normal growth conditions in crown tissue (Zhang and Shih 2007). Expression of *FaOLP2* appears to be induced in the first 12 hour following treatment with ABA or mechanical wounding, and then expression diminishes. In comparison, expression is induced within 2 hour following treatment with salicylic acid, and then *FaOLP2* RNA accumulates to a much higher level at 48 hour post treatment. These expression patterns have been interpreted as indicating the primary role for *FaOLP2* being in defense against pathogens rather than in protection against osmotic-related environmental stresses (Zhang and Shih 2007).

A gene, *FaRB7*, expressed predominately in roots was identified in an effort to obtain a promoter that would specifically target resistance genes to a region of the plant that is susceptible to vine weevil larvae and soilborne fungi (Vaughan et al. 2006). The function of *FaRB7* remains unknown, although it does contain motifs characteristic of tonoplast intrinsic proteins. More importantly, a heterologous system, the promoter region of this gene could drive expression of a reporter gene in a near root specific manner, at a level comparable to that of the cauliflower mosaic virus (CaMV) 35S promoter (Vaughan et al. 2006).

4.3.5.4 Emerging Functional Genomics

A comparative analysis of *Fragaria* dbEST revealed the existence of a number of genes in strawberry that have no known protein sequence homology outside of *Fragaria* or other members of the Rosaceae (Folta and Davis 2007). The function(s) of these sequences is particularly intriguing in that it is these sequences that have the potential to encode proteins that make *Fragaria* different from other plants. For example, these could include genes that regulate whether the stem tip swells to form an edible structure as occurs in strawberry, or recedes and retracts away from the developing fruit, as it does in the closely related *Rubus* species. Although *Fragaria* transformation is now fairly rapid and efficient, one approach to functionally characterizing these genes has been to overexpress them in *Arabidopsis* (Folta and Davis 2007), because this model plant is relatively easy to transform by dipping the inflorescence in an *Agrobacterium* culture containing the gene of interest (see, Zhang et al. 2006). Changes in rosette morphology, aberrant hypocotyl elongation in white light, severe or moderated developmental abnormalities including dwarfness and absence of trichomes, defects in greening, early or delayed flowering, and unusual phyllotaxy were observed in the *Arabidopsis* transformants, giving clues as to the possible functions of these novel Rosaceae genes (Folta and Davis 2007). Additional studies of these genes are ongoing, using overexpression and RNAi constructs to transform both diploid and octoploid strawberry (K. Folta and T. Davis, pers. comm.).

4.3.5.5 Application of Functional Genomics in Genomics-Assisted Breeding

Particularly exciting in the near future will be the ability to vastly increase our knowledge of the strawberry transcriptome, and especially the discovery of rare cell-type-specific transcripts and their expression patterns using laser-capture microdissection (LCM) and massively parallel sequencing. More than 261,000 ESTs from the shoot apical meristem of maize were obtained in a single sequencing run (Emrich et al. 2007), and using conservative alignment criteria at least 30% of these ESTs did not align to the already large number of existing maize ESTs. However, the information about strawberry gene and/or protein function that results from the use of high throughput technologies may not always be easily translated into useful ways to improve plant productivity in a straightforward manner. The problem is simpler if improvement can easily be linked to the presence or absence of a single gene or to a specific allele, however this is most frequently not the case as many traits are the result of complex, polygenic interactions.

Qin et al. (2008) have recently reviewed the transgenic strategies that have been applied to strawberry cultivar improvement including the now infamous attempt to improve freezing tolerance by transformation with an antifreeze protein gene from winter flounder (Firsov and Dolgov 1999). Molecular techniques will allow us to identify more *Fragaria* transcription factors, and we can utilize gene silencing techniques to study their function, but until genetically modified fruit is widely accepted by the consumer, the only way to improve strawberry cultivars will be through breeding. Marker-assisted selection requires that desired traits can be genetically linked to a specific molecular sequence. This is not a trivial task for octoploid strawberry, and to date, no trait-linked molecular markers are publicly available for commercial strawberry.

4.3.6 Proteomics

Proteomic analysis is dependent on being able to adequately separate large numbers of proteins in a condition suitable for mass spectral analysis, and on then being able to match the sequence of resulting peptide fragments to sequences in sequence databases. The high degree of protein sequence conservation among the members of the Rosaceae family, and the anticipated genome sequences for strawberry, peach and apple, make it much more likely that a sequence match for a given strawberry peptide will be found in sequence databases. Thus, it is becoming more and more attractive to do high throughput strawberry proteomics. Proteomic analysis substantially expands on the transcriptome analysis of gene expression to the various modified forms that a single protein may have in a cell at a single time point, or over the course of development or a response to the environment. Such modifications may affect a protein's activity, its localization, or its ability to interact with other proteins, thus extending the functionality of a given gene. Well known post-translational modifications include glycosylation, phosphorylation, ubiquitination, and lipidation such a prenylation or farnesylation. Less well known examples of protein modification include a strawberry fruit protein modified with the plant hormone auxin (Park et al. 2006).

Several approaches can be used for proteome analyses. The most common approach first separates proteins by two-dimensional gel electrophoresis (2DE). Protein spots are cut out individually, digested, and specific peptide fragments are analyzed with a mass spectrometer, typically using Matrix Assisted Laser Desorption/Ionization (MALDI)—Time of Flight (TOF) spectrometry to produce a peptide mass fingerprint, or electrospray tandem mass spectrometry (ESI-MS/MS) to obtain peptide sequences. The raw data is then processed using database searching algorithms to identify the proteins. A modification of this approach, 2-D Fluorescence Difference Gel Electrophoresis or DIGE, allows comparison of protein expression in up to three samples on the same gel by differential labeling of the samples with fluorescent dyes that are easily separated optically, but which do not affect the migration of a protein during electrophoresis.

The primary separation of protein mixtures by electrophoresis limits the analysis because it can exclude specific classes of protein, it cannot detect low abundance proteins without preliminary handling of the sample, and a single spot on a gel could contain a number of different proteins. To overcome these problems, multidimensional protein identification technology (MuDPIT), couples 2D-liquid chromatography to an electrospray ionization ion-trap mass spectrometer to resolve and identify peptides from complex mixtures, however the instrumentation may not always be available at every institution. To date, only electrophoresis has been documented for *Fragaria* proteome analysis (Alm et al. 2007; Hjernø et al. 2006; Zheng et al. 2007).

Plant proteins in general, and fruit tissue proteins in particular, are considered to be difficult to separate by 2DE because of interfering substance such as polyphenolics, terpenes, lipids, pigments, and polysaccharides that interfere with extractions and electrophoresis. Two extraction protocols were evaluated for obtaining strawberry fruit proteins for 2DE/MS (Zheng et al. 2007). From red ripe whole fruit, over 1,300 spots were separated using a phenol-based extraction protocol, and over 956 spots were separated using a hot SDS extraction followed by an additional clean-up step. In this case, peptides from trypsinized spots were first separated by HPLC, and then analyzed on a hybrid triple-quadrupole linear ion trap equipped with a nanospray ion source. Differences of molecular weight and p*I* in spot distribution between the two protocols were found in this study, and some spots were found with each protocol that were not present using the other protocol, demonstrating the need for caution when interpreting proteomic profiles of complex tissues. Differentially separated spots included a 48 kDa glycoprotein precursor, a putative HSP90, a putative 26S proteasome regulatory subunit and a small HSP.

Hjernø et al. (2006) were able to identify 58% of the *F. × ananassa* fruit proteins separated by 2D using MALDI-TOF. In some cases, N-terminal and lysine modification was required for clear spot assignment. *F. × ananassa* cv. Elsanta fruit ESTs from Plant Research International were included in their sequence database searches. They obtained partial sequence of a strawberry allergen, which is 54% homologous to the major birch pollen allergen Bet v 1, and 77% homologous to the major apple allergen. DIGE was then used to compare the protein expression in four red cultivars and four white varieties (one named cultivar, Chili Manzanar Alto), in an effort to determine why white strawberries can be tolerated by individuals with strawberry allergy. Ninety-six differentially expressed proteins were detected, and from these, 32 different proteins were identified. As expected, enzymes in the flavonoid biosynthesis pathway were expressed in red fruit and down-regulated in white. Spots that were up-regulated in white fruit included stress related proteins such as proteasome subunits, ascorbate peroxidase, and methionine sulfoxide. All four spots in red fruit identified as the Bet v 1-homologous allergen were highly down regulated in white fruit.

DIGE was also used to examine the variability in allergen content within and among strawberry cultivars (Alm et al. 2007). The results showed that for 154 different proteins, the biological variation due to growth conditions was somewhat greater than the variation among cultivars. While variation was found in allergen content among the broad spectrum of unrelated red cultivars tested, none had as low an allergen content as the three tested white varieties. It is particularly noteworthy that allergen content also varied considerably among samples of the same red variety grown under various conditions similar to those used for commercial strawberry production in southern Sweden (Alm et al. 2007).

4.3.7 Metabolomics

High throughput metabolite profiling and analysis is an integral part of the "omics" of systems biology. Metabolomics, the global analysis of cellular metabolites, can be applied to compare a specific group of cells or tissues from one genotype to another, or to determine and compare the metabolites produced by an organism in response to some stimulus or developmental state. The development of high throughput metabolic profiling useful for analysis of fruit under a very wide variety of conditions will be an important component of understanding such complex traits as flavor and aroma, production of health promoting compounds, and production of phytochemicals useful for disease prevention and treatment.

Complex parameters related to the metabolome of strawberry fruit that breeders consider include individual consumer preference and an individual's ability to detect specific compounds that contribute to strawberry flavor and attractiveness. The compounds in strawberry that play a role in the flavor and aroma of the berry have been studied for quite some time, and about 300 such compounds have been detected in ripening fruit (Zabetakis and Holden 1997). Given that the environmental conditions during growth of the fruit as well as the genetics of the parent plant contribute to the quantity and variety of compounds produced (Del Pozo-Insfran et al. 2006; Talcott 2007; Wang 2007; Wang and Lewers 2007), it is imperative to carefully assess these parameters when analyzing and comparing the metabolome of one fruit with various other cultivars or species.

The first large nontargeted metabolomic study of strawberry fruit development was reported in 2002 (Aharoni et al. 2002b). Metabolites accumulating in four stages of fruit development were analyzed using a high capacity Fourier Transform Ion Cyclotron Mass Spectrometer (FTMS). Two different extraction procedures and four modes of ionization were used, resulting in detection of over 5,000 unique, singly charged, masses of which more than half could be assigned a single empirical formula. The assigned compounds could then be identified by comparison with a commercial database of natural products. While this mode of analysis allowed for very high mass resolution, accuracy, and reproducibility, the cost, and thus the availability, of the instrumentation prevent this approach from being widely used.

FTMS was also used to broadly screen for metabolic alterations in transgenic tobacco over-expressing *FaMYB1*, a strawberry transcription factor gene that had previously been shown to suppress anthocyanin and flavonol accumulation in the transgenic tobacco (Aharoni et al. 2002b). Nine masses that were significantly different could very rapidly be detected between the transgenic and control plants, demonstrating the utility of this method for high-throughput screening of plant mutant collections, if the instrumentation is available.

A new metabolomic study of the metabolic networks in achenes as well as receptacles throughout development utilizing gas chromatography (GC) or ultra performance liquid chromatography (UPLC) in combination with more commonly available mass spectrometers has been reported (Fait et al. 2008). The Q-Tof mass spectrometer used is a hybrid quadrupole time of flight mass spectrometer with MS-MS capability that has high sensitivity, resolution, and mass accuracy. This allows exact mass measurements of small molecules that can lead directly to the empirical formula of the target compound.

A considerable metabolic change was evident in both achenes and receptables in fruit as they progressed from the green stages of development to the white, ripening and mature red stages (Fait et al. 2008). In the receptacles, the change in primary metabolism involved sugars and sugar phosphates, ascorbate, and the shikimate-precursor quinic acid. In contrast, changes in the achenes involved different metabolites including: specific amino acids; TCA cycle intermediates, and γ-amino butyric acid or GABA, which doubled in amount in the red stage. Larger changes were found for secondary metabolites in the receptacles than in the achenes. Secondary metabolites exhibiting variation in the receptacle across development included the expected compounds involved in flavor, aroma, and color such as anthocyanins, phenolic acids and terpenoids.

Many groups apply a more targeted metabolomic approach to measure compounds that define individual pathways, or specific classes of compounds (e.g., Almeida et al. 2007). Such targeted, or substrate, metabolomics rely more on extraction and chromatography to separate compounds, and may use chromatography with known standards or mass spectrometry to identify individual metabolites.

Within the last 10 years, microarray technology has been combined with a metabolomic approach to relate strawberry volatile components to the genome of the plant. An enzyme catalyzing the formation of volatile esters in strawberry fruit, SAAT, was initially identified using a cDNA microarray containing sequences from red ripe fruit (Aharoni et al. 2000). The same enzymatic activity was identified in another microarray study that combined fruit gene expression with mass spectral analysis of headspace volatiles (Carbone et al. 2006). Interestingly, the preferred substrate for the recombinant SAAT protein, octyl pentanoate, has not been found in strawberry fruit, suggesting that ester formation in strawberry is determined in large part by substrate supply and that an alcohol acyltranferase involved in fruit flavor and aroma may not have tight substrate specificity (Schwab 2003).

The ability of convergent microarray and metabolic profiling data to suggest a more comprehensive picture of fruit metabolism is also illustrated in analysis of the phenylpropanoid family, which is responsible for the red fruit coloration (Aharoni et al. 2002b; Almeida et al. 2007). Here, the metabolic profiling, together with comprehensive gene expression, and recombinant enzyme activity data combine to show that proanthocyanidins accumulate early in fruit development, and then decrease in abundance as massive synthesis of anthocyanins takes place during ripening. The data also suggest multiple roles for dihydroflavanol 4-reductase, dependent on the stage of fruit development, and clearly shows the accumulation of intermediates in monolignol biosynthesis during ripening, corroborating the lignin production in the vasculature in ripe fruit suggested by the microarray data (Aharoni et al. 2002a).

Sugars in fruit are a part of the ripening metabolome that are also commercially important, not only for the sweetness but as precursors for flavor compounds (Zabetakis and Holden 1997) and as a balance for fruit acids. Sugars in the receptacle exhibited the greatest amount of variability during development, with sucrose, glucose and fructose, which are the major soluble sugars, increasing during development (Fait et al. 2008). Sugars and storage compounds such as raffinose, isomaltose, sorbitol, glycerol and gentibiose were found to increase as achenes matured (Fait et al. 2008).

Antioxidants and anticancer agents such as ascorbic acid and ellagic acid can be especially high in strawberry (da Silva Pinto et al. 2008; Howard and Hager 2007; Wang and Lewers 2007). Resveratrol, a polyphenolic compound with health promoting properties found in grapes and grape products, has been found to occur in strawberries, albeit at lower levels than in grape (Wang et al. 2007). More resveratrol was found in achenes than in a fruit pulp, and the amounts increased as both tissues matured (Wang et al. 2007). Fait et al. (2008) discovered several peaks with ellagitannin characteristics that had not been found previously in achenes, and report that about 10 additional ellagitannin compounds may also be present. Strawberry purees enriched with achenes had higher levels of ellagic acid derivatives, as well as other polyphenols (Aaby et al. 2007), and the presence of achenes in the purees was found to increase the stability of many phenolics compounds during processing and storage.

Clearly much work remains to be done yet to integrate our understanding of production of useful compounds in strawberry. Newly developed methods and widely available instrumentation for metabolomic profiling are helping us gain an understanding of what it will take to manipulate the levels of such compounds accumulating in fruit, and will ultimately provide a useful guide to improved cultivar development.

4.3.8 Bioinformatics

Bioinformatics for *Fragaria* is essentially in its infancy, although the forthcoming genome sequences for three members of the Rosaceae will necessitate the development of a site for storing and disseminating results from comparative genomics and genome sequence annotation. No public gene expression database is currently available.

A recently upgraded (Jung et al. 2008) curated and integrated web-based relational database, The Genome Database for Rosaceae (GDR) was developed as a repository for information pertaining to Rosaceae genomics and genetics, and as a means to disseminate this information to the Rosaceae community. The GDR website has annotated strawberry EST sequences and genetic maps, as well as pages for the Rosaceae and *Fragaria* unigene projects (Table 4.3-3). Users can search ESTs or contigs, and can search the entire Rosaceae set or just the genus of interest. ESTs can be searched by several criteria, including name, sequence feature such as SSR, or GO term. For *Fragaria*, an EST details page contains links to library details, unigene information, sequence homology, and SSR/ORF information. It is anticipated that, as funding becomes available, the website will be updated with current strawberry linkage maps. The website for the GDR is: *http://www.rosaceae.org.*

Additional tools for *Fragaria* researchers include the TIGR transcript assemblies found at: *http://plantta.jcvi.org/* (Childs et al. 2007). These assemblies represent genes that are known to be expressed in a plant and are constructed from ESTs and entries in the NCBI GenBank nucleotide database (full-length and partial cDNAs), but exclude sequences derived from whole genome annotation projects. Separate assemblies were constructed for *F. x ananassa* and *F. vesca* (Table 4.3-3), and the assemblies are searchable by name, gene annotation, or by sequence similarity. The entire assembly can be downloaded from the site.

4.3.9 Future Prospects

4.3.9.1 Potential for Expansion of Productivity

Several avenues are open for expanding productivity and consumption of strawberries. Among these would be the development of varieties that expand the growing season; for example, plants capable of producing throughout the hot summer months, or that can withstand late frosts during flowering and fruit development. Development of plants that are more resistant to biotic stresses, especially the fungi that cause gray mold and anthracnose, would greatly decrease the amount of fruit that is lost both in the field and postharvest. Another area for expansion of production would be the improvement of existing native species, such as the large, white or pale pink fruited, cultivated form of *F. chiloensis* (see, Carrasco et al. 2007). Each of these avenues is dependent on the further development of molecular markers for more rapid success in breeding programs, and/ or on the development of efficient and accepted methods for transferring desired genes into the crop.

4.3.9.2 Genetic Manipulation

The availability of transformation systems for strawberry (Folta and Davis 2007; Folta and Dhingra 2007; Slovin et al. 2009) has made it possible to introduce heterologous genes that have the potential to improve fruit quality, or resistance to biotic and abiotic stresses. *Agrobacterium*-mediated transformation was used to introduce a bean chitanase gene (Vellicce et al. 2006) or a thaumatin gene from *Thaumatococcus daniellii* (Schestibratov and Dolgov 2006) to produce plants that are more resistant to gray mold *Botrytis cinerea*. Introduction of a tobacco osmotin gene into strawberry resulted in plants that appear more salt tolerant in laboratory assays (Husaini and Abdin 2008) and introduction of an acidic dehydrin gene from wheat improved freezing tolerance in leaves (Houde et al. 2004).

The ultimate tests of the efficacy of such genetic manipulations with heterologous genes are the resultant yield and quality of the fruit produced in commercial situations, however, the current level of acceptance of genetically modified fruit has hampered interest in doing field trials. A transformation system for generating transgenic plants without the need for antibiotic resistance or other selectable markers, or for crossing out selectable marker genes after transformation, has been reported for use in strawberry (Schaart et al. 2004). Combined with strawberry tissue-specific promoters (e.g., Vaughan et al. 2006), such systems may help to increase consumer acceptance of genetically modified fruit.

Despite the low level of consumer approval at the present time, there is continuing interest in developing efficient methods for genetic engineering of strawberry, as evidenced by the report on the production of hygromycin-resistant plants in temporary immersion bioreactors (TIB) (Hanhineva and Kärenlampi 2007). In this apparently less labor intensive system, shoot clusters emerged after three weeks from callus formed on leaf pieces in the small TIB containers, in which plant tissues are immersed for ten seconds every four hours in liquid medium.

Development of a floral dip method for rapid transformation such as that used for *Arabidopsis* (Zhang et al. 2006) would be a substantial improvement on tissue culture methods for functional analysis of genes and their promoters in strawberry. In *Arabidopsis*, the *Agrobacterium* needs to be delivered to the ovules for efficient transformation to occur (Desfeux et al. 2000). Floral-dip transformation of *Capsella bursa-pastoris* (Bartholmes et al. 2008), a close relative of *Arabidopsis*, appears to also occur when the ovules have already formed. The mechanism for transformation of radish (Curtis and Nam 2001), another relative of *Arabidopsis*, appears to be different. Anecdotal accounts exist of unsuccessful attempts to use floral dip to transform strawberry, however, no systematic study of developmental stage of inflorescence, detergent type or concentration, or selection method for floral dip transformation has yet been documented.

4.3.9.3 Increasing Consumption

The growing consumer awareness of the health benefits of eating fruits and vegetables could lead to an overall increase in fruit consumption, as well as a demand for value added products with increased amounts of phytonutriceuticals. Strawberry phenolic compounds reportedly have a range of possible anti-cancer and anti-heart disease properties (Howard and Hager 2007), however very little is known about how the levels of most of these, such as ellagic acid for example, are determined by the genotype or the environment of the plant. Clearly, work done at the molecular level in the area of strawberries and human health has potential to lead to an increase in fruit consumption.

4.3.9.4 The Fragaria Research Community

The next few years should see a dramatic increase in the use of molecular tools in the process of development of new and better strawberry cultivars. The availability of sequence data for members of the Rosaceae family, together with more populated strawberry genetic maps and functional genomics tools such as the diploid strawberry, make this a very exciting time for *Fragaria* research. Although there are bound to be limitations on information sharing because strawberry production is a commercial enterprise, major improvements in strawberries for the future will ultimately come from sharing information that increases an overall understanding of the molecular biology of this very interesting plant.

References

Aaby K, Wrolstad RE, Ekeberg D, Skrede G (2007) Polyphenol composition and antioxidant activity in strawberry purees; impact of achene level and storage. J Agric Food Chem 55: 5156–5166.

Agius F, González-Lamothe R, Caballero JL, Muñoz-Blanco J, Botella MA, Valpuesta V (2003) Engineering increased vitamin C levels in plants by overexpression of a D-galacturonic acid reductase. Nature Biotech 21: 177–181.

Aharoni A, Keizer LCP, Bouwmeester HJ, Sun Z, Alvarez-Huerta M, Verhoeven HA, Blaas J, van Houwelingen AMML, De Vos RCH, van der Voet H, Jansen RC, Guis M, Mol J, Davis RW, Schena M, van Tunen AJ, O'Connell AP (2000) Identification of the SAAT gene involved in strawberry flavor biogenesis by use of DNA microarrays. The Plant Cell 12: 647–661.

Aharoni A, Keizer LCP, Van Den Broeck HC, Blanco-Portales R, Munoz-Blanco J, Bois G, Smit P, De Vos RCH, O'Connell AP (2002a) Novel insight into vascular, stress, and auxin-dependent and -independent gene expression programs in Strawberry, a non-climacteric fruit. Plant Phys 129: 1019–1031.

Aharoni A, Ric de Vos CH, Verhoeven HA, Maliepaard CA, Kruppa G, Bino R, Goodenowe DB (2002b) Nontargeted metabolome analysis by use of Fourier Transform Ion Cyclotron Mass Spectrometry. OMICS. J Int Biology 6: 217–234.

Aharoni A, Giri AP, Verstappen FWA, Bertea C, M. , Sevenier R, Sun Z, Jongsma MA, Schwab W, Bouwmeester HJ (2004) Gain and loss of fruit flavor compounds produced by wild and cultivated strawberry species. The Plant Cell 16: 3110–3131.

Albani MC, Battey NH, Wilkinson MJ (2004) The development of ISSR-derived SCAR markers around the *SEASONAL FLOWERING LOCUS* (*SFL*) in *Fragaria vesca*. Theor Appl Genet.

Alm R, Ekefjärd A, Krogh M, Häkkinen J, Emanuelsson C (2007) Proteomic variation Is as large within as between strawberry varieties. J Proteome Res 6: 3011–3020.

Almeida JRM, D'Amico E, Preuss A, Carbone F, de Vos CHR, Deiml B, Mourgues F, Perrotta G, Fischer TC, Bovy AG, Martens S, Rosati C (2007) Characterization of major enzymes and genes involved in flavonoid and proanthocyanidin biosynthesis during fruit development in strawberry (*Fragaria* x *ananassa*). Arch Biochem Biophys 465: 61–71.

Archbold DD, Dennis FG (1984) Quantification of free ABA and free and conjugated IAA in strawberry achene and receptacle tissue during fruit development. J Am Soc Hort Sci 109: 330–335.

Bartholmes C, Nutt P, Theien G (2008) Germline transformation of Shepherd's purse (*Capsella bursa-pastoris*) by the 'floral dip' method as a tool for evolutionary and developmental biology. Gene 409: 11–19.

Bassil NV, Gunn M, Folta KM, Lewers KS (2006a) Microsatellite markers for *Fragaria* from "Strawberry Festival" expressed sequence tags. Mol Ecol Notes 6: 473–476

Bassil NV, Njuguna W, Slovin JP (2006b) EST-SSR markers from *Fragaria vesca* L. cv. Yellow Wonder. Mol Ecol Notes 6: 806–809.

Battey NH, LeMiere P, Tehranifar A, Cekic C, Taylor S, Shrives KJ, Hadley P, Greenland AJ, Darby J, Wilkinson MJ (1998) Genetic and environmental control of flowering in strawberry. In: KE Cockshull, D Gray, GB Seymour, B Thomas (eds) Genetic and Environmental Manipulation of Horticultural Crops. CABI Publishing, New York, pp 111–131.

Biswas MK, Islam R, Hossain M (2007) Somatic embryogenesis in strawberry (*Fragaria* sp.) through callus culture. Plant Cell Tiss Org Cult 90: 49–54.

Carbone F, Mourgues F, Biasioli F, Gasperi F, Mark TD, Rosati C, Perrotta G (2006) Development of molecular and biochemical tools to investigate fruit quality traits in strawberry elite genotypes. Mol Breed 18: 127–142.

Carrasco B, Garcés M, Rojas P, Saud G, Herrera R, Retamales JB, Caligari PDS (2007) The Chilean strawberry [*Fragaria chiloensis* (L.) Duch.]: genetic diversity and structure. J Am SocHort Sci 132: 501–506.

Casado-Diaz A, Encinas-Villarejo S, Santos Bdl, Schiliro E, Yubero-Serrano E-M, Amil-Ruiz F, Pocovi MI, Pliego-Alfaro F, Dorado G, Rey M, Romero F, Munoz-Blanco J, Caballero J-L (2006) Analysis of strawberry genes differentially expressed in response to *Colletotrichum* infection. Physiol Plantarum 128: 633–650.

Cekic C, Battey NH, Wilkinson MJ (2001) The potential of ISSR-PCR primer-pair combinations for genetic linkage analysis using the seasonal flowering locus in *Fragaria* as a model. Theor Appl Genet 103: 540–546.

Childs KL, Hamilton JP, Zhu W, Ly E, Cheung F, Wu H, Rabinowicz PD, Town CD, Buell CR, Chan AP (2007) The TIGR plant transcript assemblies database. Nucleic Acids Res 35: D846–D851.

Cordero de Mesa M, Jimnez-Bermudez S, Pliego-Alfaro F, Quesada MA, Mercado JA (2000) *Agrobacterium* cells as microprojectile coating: a novel approach to enhance stable transformation rates in strawberry. Funct Plant Biol 27: 1093–1100.

Curtis IS, Nam HG (2001) Transgenic radish (*Raphanus sativus* L. *longipinnatus* Bailey) by floral-dip method—plant development and surfactant are important in optimizing transformation efficiency. Transgenic Res 10: 363–371.

da Silva Pinto M, Lajolo FM, Genovese MI (2008) Bioactive compounds and quantification of total ellagic acid in strawberries (*Fragaria* x *ananassa* Duch.). Food Chem 107: 1629–1635.

Darnell RL, Cantliffe DJ, Kirschbaum DS, Chandler CK (2003) The physiology of flowering in strawberry. In: J Janick (ed) Horticultural Reviews. John Wiley & Sons, Inc., pp 325–349.

Davis TM, Pollard JE (1991) *Fragaria vesca* chlorophyll mutants. HortScience 26: 311.

Davis TM, Folta KM, Shields M, Zhang Q (2007) Gene pair markers: an innovative tool for comparative linkage mapping. In: F Takeda (ed) 6th North Amer Strawberry Symp. Amer Soc Hort Sci, Ventura CA, pp 105–107.

Davis TM, Folta KM, Bennetzen JL, Shields M, Zhang Q, Pontaroli A, Tombolato D, DiMeglio L (2008a) The *Fragaria* genome: structure, organization, and transmission. Abstracts, Fourth International Rosaceae Genomics Conference, Pucon, Chile, p 73.

Davis TM, Shields M, Zhang Q, Poulsen EG, Folta KM, Bennetzen JL, San Miguel P (2008b) The strawberry genome is coming into view. Acta Horticulturae (ISHS) 842: 533–536.

de la Fuente JI, Amaya I, Castillejo C, Sanchez-Sevilla JF, Quesada MA, Botella MA, Valpuesta V (2006) The strawberry gene FaGAST affects plant growth through inhibition of cell elongation. J Exp Bot 57: 2401–2411.

Debnath SC, Teixeira da Silva JA (2007) Strawberry culture *in vitro*: applications in genetic transformation and biotechnology. Fruit, Vegetable, and Cereal Science Biotechnology 1: 1–2.

Del Pozo-Insfran D, Duncan CE, Yu KC, Talcott ST, Chandler CK (2006) Polyphenolics, ascorbic acid, and soluble solids concentrations of strawberry cultivars and selections grown in a winter annual hill production system. J Amer Soc Hort Sci 131: 89–96.

Desfeux C, Clough SJ, Bent AF (2000) Female reproductive tissues are the primary target of *Agrobacterium*-mediated transformation by the *Arabidopsis* floral-dip method. Plant Physiol 123: 895–904.

Durner EF, Poling EB (1985) Comparison of three methods for determining the floral or vegetative status of strawberry plants. J Amer Soc Hort Sci 110: 808–811.

Emrich SJ, Barbazuk WB, Li L, Schnable PS (2007) Gene discovery and annotation using LCM-454 transcriptome sequencing. Genome Res 17: 69–73.

Fait A, Hanhineva K, Beleggia R, Dai N, Rogachev I, Nikiforova VJ, Fernie AR, Aharoni A (2008) Reconfiguration of the achene and receptacle metabolic networks during strawberry fruit development. Plant Physiol 148: 730–750.

Firsov AP, Dolgov SV (1999) Agrobacterial transformation and transfer of the antifreeze protein gene of winter flounder to the strawberry. Acta Hortaculturae (ISHS) 484: 581–586.

Folta KM, Davis TM (2006) Strawberry genes and genomics. Current Reviews in Plant Sciences 25: 399–415.

Folta KM, Davis TM (2007) Transformation systems to study gene function in *Fragaria*. In: F Takeda (ed) 6th N Amer Strawberry Symp. Amer Soc Hort Sci, Ventura, CA, pp 94–99

Folta KM, Dhingra A (2007) Transformation of strawberry: the basis for translational genomics in *Rosaceae*. In Vitro Cell Dev Biol Plant 42: 482–490.

Folta KM, Dhingra A, Howard L, Stewart PJ, Chandler CK (2006) Characterization of LF9, an octoploid strawberry genotype selected for rapid regeneration and transformation. Planta 224: 1058–1067.

Garcia EL, Barrett DM (2000) Preservative treatments for fresh-cut fruits and vegetables. In: O Lamikanra (ed) Fresh-cut Fruits and Vegetables Science, Technology and Market. CRC Press, Boca Raton, pp 267–303.

Gil-Ariza DJ, Amaya I, Botella MA, Munoz-Blanco J, Caballero JL, Lopez-Aranda JM, Valpuesta V, Sanchez-Sevilla JF (2006) EST-derived polymorphic microsatellites from cultivated strawberry (*Fragaria x ananassa*) are useful for diversity studies and varietal identification among *Fragaria* species. Mol Ecol Notes 4: 366–368.

Given NK, Venis MA, Grierson D (1988) Hormonal regulation of ripening in the strawberry, a non-climacteric fruit. Planta 174: 402–406.

Griesser M, Hoffmann T, Bellido ML, Rosati C, Fink B, Kurtzer R, Aharoni A, Munoz-Blanco J, Schwab W (2008a) Redirection of flavonoid biosynthesis through the down-regulation of an anthocyanidin glucosyltransferase in ripening strawberry fruit. Plant Physiol 146: 1528–1539.

Griesser M, Vitzthum F, Fink B, Bellido ML, Raasch C, Munoz-Blanco J, Schwab W (2008b) Multi-substrate flavonol O-glucosyltransferases from strawberry (Fragariaxananassa) achene and receptacle. J Exp Bot 59: 2611–2625.

Hanhineva KJ, Kärenlampi SO (2007) Production of transgenic strawberries by temporary immersion bioreactor system and verification by TAIL-PCR. BMC Biotechnology 7: 11.

Herman EM, Helm RM, Jung R, Kinney AJ (2003) Genetic modification removes an immunodominant allergen from soybean. Plant Physiol 132: 36–43.

Hjernø K, Alm R, Canbäck B, Matthiesen R, Trajkovski K, Björk L, Roepstorff P, Emanuelsson C (2006) Down-regulation of the strawberry Bet v 1-homologous allergen in concert with the flavonoid biosynthesis pathway in colorless strawberry mutant. Proteomics 6: 1574–1587.

Hoffmann T, Kalinowski G, Schwab W (2006) RNAi-induced silencing of gene expression in strawberry fruit (*Fragaria* x *ananassa*) by agroinfiltration: a rapid assay for gene function analysis. The Plant J 48: 818–826.

Hokanson SC, Maas JL (2001) Strawberry biotechnology. In: J Janick (ed) Plant Breeding Reviews. John Wiley & Sons, Inc., New York, pp 139–180.

Houde M, Dallaire S, N'Dong D, Sarhan F (2004) Overexpression of the acidic dehydrin WCOR410 improves freezing tolerance in transgenic strawberry leaves. Plant Biotech J. 2: 381–387.

Howard LR, Hager TJ (2007) Berry fruit phytochemicals. In: Y Zhao (ed) Berry Fruit: Value-Added Products for Health Promotion. CRC Press, Boca Raton, pp 73–104.

Husaini AM, Abdin MZ (2008) Development of transgenic strawberry (*Fragaria* x *ananassa* Duch.) plants tolerant to salt stress. Plant Science 174: 446–455.

Hytönen T, Elomaa P, Moritz T, Junttila O (2009) Gibberellin mediates daylength-controlled differentiation of vegetative meristems in strawberry (*Fragaria xananassa* Duch) BMC Plant Biology 9: 18.

Jiménez-Bermudez S, Redondo-Nevado J, Munoz-Blanco J, Caballero JL, Lopez-Aranda JM, Valpuesta V, Pliego-Alfaro F, Quesada MA, Mercado JA (2002) Manipulation of strawberry fruit softening by antisense expression of a pectate lyase gene. Plant Physiol 128: 751–759.

Jung S, Staton M, Lee T, Blenda A, Svancara R, Abbott A, Main D (2008) GDR (Genome Database for Rosaceae): integrated web-database for Rosaceae genomics and genetics data. Nucl Acids Res 36: D1034–1040.

Kaur S, Marija T, Cogan NOI, Spanenberg GC, Forster JW (2008) In vitro SNP discovery in the cultivated octoploid strawberry (*Fragaria* x *ananassa*). Plant and Animal Genome XVI, San Diego, CA.

Keniry A, Hopkins CJ, Jewell E, Morrison B, Spangenberg GC, Edwards D, Batley J (2006) Identification and characterization of simple sequence repeat (SSR) markers from *Fragaria xananassa* expressed sequences. Mol Ecol Notes 6: 319–322.

Kim YJ, Lin NC, Martin GB (2002) Two distinct *Pseudomonas* effector proteins interact with the Pto kinase and activate plant immunity. Cell 109: 598–598.

Kurokura T, Inaba Y, Sugiyama N (2006) Histone H4 gene expression and morphological changes on shoot apices of strawberry (*Fragaria* x *ananassa* Duch.) during floral induction Scientia Hort 110: 192–197.

Landmann C, Fink B, Schwab W (2007) FaGT2: a multifunctional enzyme from strawberry (*Fragaria* × *ananassa*) fruits involved in the metabolism of natural and xenobiotic compounds. Planta 226: 417–428.

Lewers KS, Styan SMN, Hokanson SC (2005) Strawberry GenBank-derived and genomic simple sequence repeat (SSR) markers and their utility with strawberry, blackberry, and red and black raspberry. J Am Soc Hort Sci 130: 102–115.

Lurie S (1998) Postharvest heat treatments. Postharvest Biology and Technology 14: 257–269

Maas JL (ed) (1984) Compendium of Strawberry Diseases. APS Press, St. Paul.

Martínez GA, Civello PM (2008) Effect of heat treatments on gene expression and enzyme activities associated to cell wall degradation in strawberry fruit. Postharvest Biol Tech 49: 38–45.

Martinez Zamora MC, Castagnaro AP, Dias Ricci JC (2004) Isolation and diversity analysis of resistance gene analogs (RGAs) from cultivated and wild strawberries. Molecular and General Genetics 272: 480–487.

Martinez Zamora MC, Castagnaro AP, Dias Ricci JC (2008) Genetic diversity of Pto-like serine/threonine kinase disease resistance genes in cultivated and wild strawberries. Journal of Molecular Evolution 67: 211–217.

Mercado JA, Martín-Pizarro C, Pascual L, Quesada MA, Pliego-Alfaro F, de los Santos B, Romero F, Galvez J, Rey M, de la Viña G, Llobell A, Yubero-Serrano E-M, Muñoz-Blanco J, Caballero JL (2007) Evaluation of tolerance of *Colletotrichum acutatum* in strawberry plants transformed with *Trichoderma*-derived genes. Acta Hort (ISHS) 738: 383–388.

Mezzetti B, Constantini E (2006) Strawberry (*F.* × *ananassa*). Meth Mol Biol 344: 287–295.

Mouhu K, Hytönen T, Folta K, Rantanen M, Paulin L, Auvinen L, Elomaa P (2009) Identification of flowering genes in strawberry, a perennial SD plant. BMC Plant Biology 9: 122.

Musidlowska-Persson A, Alm R, Emanuelsson C (2007) Cloning and sequencing of the Bet v 1-homologous allergen Fra a 1 in strawberry (*Fragaria ananassa*) shows the presence of an intron and little variability in amino acid sequence. Mol Immunol 44: 1245–1252.

Mut P, Bustamante C, Martinez G, Alleva K, Sutka M, Civello M, Amodeo G (2008) A fruit-specific plasma membrane aquaporin subtype PIP1; 1 is regulated during strawberry (*Fragaria* x *ananassa*) fruit ripening. Physiol Plantarum 132: 538–551.

Oosumi T, Gruszewski HA, Blischak LA, Baxter AJ, Wadl PA, Shuman JL, Veilleux RE, Shulaev V (2006) High-efficiency transformation of the diploid strawberry (*Fragaria vesca*) for functional genomics. Planta 223: 1219–1230.

Osorio S, Castillejo C, Quesada MA, Medina-Escobar N, Brownsey GJ, Suau R, Heredia A, Botella MA, Valpuesta V (2008) Partial demethylation of oligogalacturonides by pectin methyl esterase 1 is required for eliciting defence responses in wild strawberry (*Fragaria vesca*). Plant J 54: 43–55.

Owen HR, Miller AR (1996) Haploid plant regeneration from other cultures of three North American cultivars of strawberry (*Fragaria* x *ananassa* Duch.). Plant Cell Rep 15: 905–909.

Palomer X, Llop-Tous I, Vendrell M, Krens FA, Schaart JG, Boone MJ, Valk Hvd, Salentijn EMJ (2006) Antisense down-regulation of strawberry *endo*-b-(1,4)-glucanase genes does not prevent fruit softening during ripening. Plant Sci 171: 640–646.

Pankov VV (1992) Effect of growth regulators on plant production of strawberry mother plants. Sciencia Hort 52: 157–161.

Park S, Cohen JD, Slovin JP (2006) Strawberry fruit protein with a novel indole-acyl modification. Planta 224: 1015–1022.

Pontaroli A, Rogers RL, Zhang Q, Sheilds M, Davis TM, Folta KM, SanMiguel P, Bennetzen JL (2009) Gene content and distribution in the nuclear genome of *Fragaria vesca*. The Plant Genome 2: 193–201.

Qin Y, Teixeira da Silva JA, Zhang L, Zhang S (2008) Transgenic strawberry: State of the art for improved traits. Biotech Adv 26: 219–232.

Raab T, Lopez-Raez JA, Klein D, Caballero JL, Moyano E, Schwab W, Munoz-Blanco J (2006) FaQR, required for the biosynthesis of the strawberry flavor compound 4-hydroxy-2,5-dimethyl-3(2H)-furanone, encodes an enone oxidoreductase. Plant Cell 18: 1023–1037.

Rousseau-Gueutin M, Lerceteau-Köhler E, Barrot L, Sargent DJ, Monfort A, Simpson D, Arús P, Guérin G, Denoyes-Rothan B (2008) Comparative genetic mapping between octoploid and diploid Fragaria species reveals a high level of colinearity between their genomes and the essential disomic behavior of the cultivated octoploid strawberry. Genetics 179: 2045–2060.

Ruiz-Rojas JJ, Pattison J, Sargent DJ, Oosumi T, Shulaev V, Veilleux RE (2008) Mapping insertional mutants of the diploid strawberry, *Fragaria vesca*, through SNP discovery in flanking regions. Abstract: Fourth International Rosaceae Genomics Conference, Pucon, Chile, p 18.

Santiago-Domenech N, Jimenez-Bemudez S, Matas AJ, Rose JKC, Munoz-Blanco J, Mercado JA, Quesada MA (2008) Antisense inhibition of a pectate lyase gene supports a role for pectin depolymerization in strawberry fruit softening. J Exp Bot 59: 2769–2779.

Sargent DJ, Battey NH, Wilkinson MJ, Simpson DW (2006) The detection of QTL associated with vegetative and reproductive traits in diploid *Fragaria*: a preliminary study. Acta Hort (ISHS) 708: 471–474.

Sargent DJ, Rys A, Nier S, Simpson DW, Tobutt KR (2007) The development and mapping of functional markers in *Fragaria* and their transferability and potential for mapping in other genera. Theor Appl Genet 114: 373–384.

Sargent DJ, Cipriani G, Vilanova S, Gil-Ariza D, Arús P, Simpson DW, Tobutt KR, Monfort A (2008) The development of a bin mapping population and the selective mapping of 103 markers in the diploid Fragaria reference map. Genome 51: 120–127.

Schaart JG, Krens FA, Pelgrom KTB, Mendes O, Rouwendal GJA (2004) Effective production of marker-free transgenic strawberry plants using inducible site-specific recombination and a bifunctional selectable marker gene. Plant Biotech J 2: 233–240.

Schestibratov KA, Dolgov SV (2006) Genetic engineering of strawberry cv. Firework and cv. Selekta for taste improvement and enhanced disease resistance by introduction of *THAUII* gene. Acta Hort (ISHS) 708: 475–482.

Schwab W, Aharoni A, Raab T, Perez AG, Sanz C (2001) Cytosolic aldolase is a ripening related enzyme in strawberry (*Fragaria xananassa*). Phytochemistry 56: 407–415.

Schwab W (2003) Metabolome diversity: too few genes, too many metabolites? Phytochemistry 62: 837–849.

Shi Y, Zhang Y, Shih DS (2006) Cloning and expression analysis of two [beta]-1,3-glucanase genes from Strawberry. J Plant Physiol 163: 956–967.

Shulaev V, Korban SS, Sosinski B, Abbott AG, Aldwinckle HS, Folta KM, Iezzoni A, Main D, Arus P, Dandekar AM, Lewers K, Brown SK, Davis TM, Gardiner SE, Potter D, Veilleux RE (2008) Multiple models for Rosaceae genomics. Plant Physiol 147: 985–1003.

Slovin J, Rabinowicz PD (2007) *Fragaria vesca*, a useful tool for Rosaceae genomics. In: F Takeda (ed) 6th North American Strawberry Symposium. American Society for Horticultural Science, Ventura, CA, pp 112–117.

Slovin JP, Schmitt K, Folta KM (2009) An inbred line of *Fragaria vesca* f. *semperflorens* for genomic and molecular genetic studies in the Rosaceae. Plant Methods 5: 15.

Smith B (2007) Developmental stage and temperature affect strawberry flower and fruit susceptibility to anthracnose. In: F Takeda , DT Handley , EB Poling (eds) 2007 North American Strawberry Symposium. North American Strawberry Growers Association, Ventura Beach, CA, pp 55–57.

Smith B (2007) Developmental stage and temperature affect strawberry flower and fruit susceptibility to anthracnose. In: F Takeda , DT Handley , EB Poling (eds) 2007 North American Strawberry Symposium. N Am Strawberry Growers Assoc, Ventura Beach, CA, pp 55–57.

Spolaore S, Trainotti L, Casadoro G (2001) A simple protocol for transient gene expression in ripe fleshy fruit mediated by *Agrobacterium*. J Exper Bot 52: 845–850.

Stewart PJ, Winslow AR, Folta KM (2007) An initial characterization of *Fragaria* CONSTANS gene structure and expression in short-day and day-neutral cultivars. In: F Takeda (ed) 6th North American Strawberry Symposium. Amer Soc Hort Sci, Ventura CA, pp 100–104.

Suarez-Lopez P, Wheatley K, Robson F, Onouchi H, Valverde F, Coupland G (2001) CONSTANS mediates between the circadian clock and the control of flowering in Arabidopsis. Nature 410: 1116–1120.

Talcott ST (2007) Chemical components of berry fruits. In: Y Zhao (ed) Berry Fruit Value-Added Products for Health Promotion. CRC Press, Boca Raton, pp 51–72.

Uratsu S, Ahmadi H, Bringhurst RS, Dandekar A (1991) Relative virulence of *Agrobacterium* strains on strawberry. HortScience 26: 196–199.

Vaughan SP, James DJ, Lindsey K, Massiah AJ (2006) Characterization of FaRB7, a near root-specific gene from strawberry (*Fragaria* x *ananassa* Duch.) and promoter activity analysis in homologous and heterologous hosts. J Exper Bot 57: 3901–3910.

Vellicce GR, Ricci JCDa, Herna´ndez Lz, Castagnaro AP (2006) Enhanced resistance to *Botrytis cinerea* mediated by the transgenic expression of the chitinase gene ch5B in strawberry. Transgenic Res 15: 57–68.

Villarreal NM, Rosli HG, Martinez GA, Civello PM (2008) Polygalacturonase activity and expression of related genes during ripening of strawberry cultivars with contrasting fruit firmness. Postharvest Biol Technol 47: 141–150.

Wallin A (1997) Somatic hybridization in *Fragaria*. Acta Hort (ISHS) 439: 63–66.

Wang J, Ge H, Peng S, Zhang H, Chen P, Xu J (2004) Transformation of strawberry (*Fragaria ananassa* Duch.) with late embryogenesis abundant protein gene. J Hort Sci Biotech 79: 735–738.

Wang SY (2007) Antioxidant capacity and phenolic content of berry fruits as affected by genotype, preharvest conditions, maturity, and postharvest handling. In: Y Zhao (ed) Berry Fruit: Value-Added Products for Health Promotion. CRC Press, Boca Raton, pp 147–186.

Wang SY, Lewers KS (2007) Antioxidant capacity and flavonoid content in wild strawberries. J Amer Soc Hort Sci 132: 629–637.

Wang SY, Chen C-T, Wang CY, Chen P (2007) Resveratrol content in strawberry fruit is affected by preharvest conditions. J Agric Food Chem 55: 8269–8274.

Zabetakis I, Holden MA (1997) Strawberry flavour: analysis and biosynthesis. J Sci Food Agri 74: 421–434.

Zhang X, Henriques R, Lin S-S, Niu Q-W, Chua N-H (2006) *Agrobacterium*-mediated transformation of *Arabidopsis thaliana* using the floral dip method. Nat Protocols 1: 641–646.

Zhang Y, Shih DS (2007) Isolation of an osmotin-like protein gene from strawberry and analysis of the response of this gene to abiotic stresses. J Plant Physiol 164: 68–77.

Zheng Q, Song J, Doncaster K, Rowland E, Byers DM (2007) Qualitative and quantitative evaluation of protein extraction protocols for Apple and Strawberry fruit suitable for two-dimensional electrophoresis and mass spectrometry analysis. J Agricultural Food Chem 55: 1663–1673.

Index

Color Plate Section

Chapter 1

Figure 1-1 Highbush blueberry fruit on the new cv. Draper.

Chapter 2

Figure 2-2 Ripe cranberry fruits showing seeds and locules.

Figure 2-4 Cranberry flowers.

Figure 2-5 A diseased cranberry plant exhibiting necrotic tissues.

Chapter 3

Figure 3-1 Common traits of interest found in blackberry. A) A terminal inflorescence of a primocane-fruiting genotype in summer, with a floricane fruit picked from the same plant. B) Long fruit of "Natchez" blackberry, a shape desired by marketers and growers. C) Poor fruit set of a primocane-fruiting genotype due to high heat during summer bloom. D) Fruit set of Prime-Jim in moderate climate during bloom in coastal California. E) Red drupe development after harvest and storage; preference is for retention of full black color. F) White drupe development, thought to be due to effects of intense ultraviolet light. G) Stem of the thorned genotype "Prime-Jim" F) Stem of the thornless genotype "Arapaho". (Photos A-F kindly provided by J. R. Clark, G-H provided by J. D. Swanson).

Chapter 4

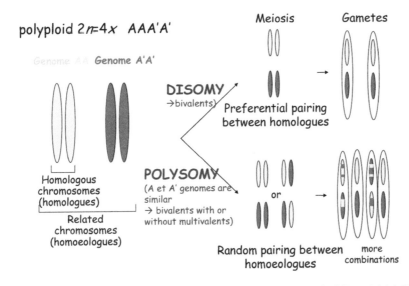

Figure 4.2-3 An example of chromosome pairing at metiosis in a tetraploid (2n = 4x) AAA′A′ under disomy and polysomy behavior.

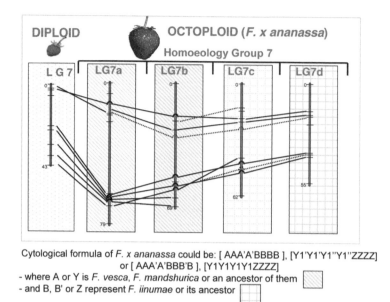

Cytological formula of *F. x ananassa* could be: [AAA'A'BBBB], [Y1'Y1'Y1''Y1''ZZZZ]
or [AAA'A'BBB'B], [Y1Y1Y1Y1ZZZZ]
- where A or Y is *F. vesca, F. mandshurica* or an ancestor of them
- and B, B' or Z represent *F. iinumae* or its ancestor

Figure 4.2-5 High levels of conserved macrosynteny and colinearity within homo(eo)logous linkage group 7 of *F. × ananassa* and between this octoploid homoeologous group and its corresponding diploid linkage group 7 from the reference diploid linkage map. Two of the four sub-genomes, which would belong to *F. vesca* or *F. mandshurica* or their ancestor, are indicated using slanting lines while the two other sub-genomes, which can be attributed to *F. iinumae* or its ancestor, are indicated using crossed lines.

T - #0025 - 160425 - C6 - 234/156/13 [15] - CB - 9781578087075 - Gloss Lamination